All the People

A H I S T O R Y O F U S

The picture on the cover, Night Game (Practice Time), *was painted by Ralph Fasanella in 1979. Fasanella, a union organizer, is considered a "primitive" or "folk" artist, which means he didn't have much artistic training. Folk artists use talent and imagination; you can see them both in this painting of the all-American game. The names on the scoreboard include Ruth, Gehrig, Robinson, and Williams. Something is wrong with that—what is it? And what are those cages for? (Note the painting's title for a clue.)*

Oxford University Press

In these books you will find explorers, farmers, cow-boys, heroes, villains, inventors, presidents, poets, pirates, artists, slaves, teachers, revolutionaries, priests, musicians— the girls and boys, men and women, who all became Americans....

OXFORD
A HISTORY OF
US
BOOK TEN

All the People

Joy Hakim

Oxford University Press
New York

Oxford University Press

Oxford   New York

Athens   Auckland   Bangkok   Bombay
Calcutta   Cape Town   Dar es Salaam   Delhi
Florence   Hong Kong   Istanbul   Karachi
Kuala Lumpur   Madras   Madrid   Melbourne
Mexico City   Nairobi   Paris   Singapore
Taipei   Tokyo   Toronto

and associated companies in
Berlin   Ibadan

Copyright © 1995 by Joy Hakim

Designer: Mervyn E. Clay

Maps copyright © 1995 by Wendy Frost and Elspeth Leacock

Produced by American Historical Publications

Published by Oxford University Press, Inc.

200 Madison Avenue, New York, New York 10016

Oxford is a registered trademark of Oxford University Press

Library of Congress Cataloging-in-Publication Data
Hakim, Joy.
All the people / Joy Hakim.
p.   cm.—(A history of US: bk. 10)
Includes bibliographical references and index.
ISBN 0-19-507763-6 (lib. ed.)—ISBN 0-19-507765-2 (series, lib. ed.)
ISBN 0-19-507764-4 (paperback ed.)—ISBN 0-19-507766-0 (series, paperback ed.)
ISBN 0-19-509515-4 (trade hardcover ed.)—ISBN 0-19-509484-0 (series, trade hardcover ed.)
1. United States—History—1945– —Juvenile literature.
[1. United States—History—1945–] I. Title. II. Series: Hakim, Joy. History of US; 10.
E178.3.H22 1995 vol. 10
[E741]
973.92—dc20   93-28564
CIP
AC

1 3 5 7 9 8 6 4 2
Printed in the United States of America
on acid-free paper

The words on page 5 (opposite) are from "This Land Is Your Land," words and music by Woody Guthrie. Copyright © 1956 (renewed), 1958 (renewed), and 1970 by Ludlow Music, New York, N.Y. Used by permission. The extracts on pages 71 and 85 are from *Colored People* by Henry Louis Gates, Jr. Copyright © 1994 by Henry Louis Gates, Jr. Reprinted by permission of Alfred A. Knopf, Inc. The extract on page 84 is from *Warriors Don't Cry* by Melba Pattillo Beals. Copyright © 1994, 1995 by Melba Beals. Reprinted by permission of Pocket Books, a division of Simon & Schuster, Inc. The poem on page 145 is "Mexico Is Sinking" by Guillermo Gomez-Peña. Copyright © 1986 by Guillermo Gomez-Peña. First published in *High Performance*, no. 35, 1986. Reprinted by permission of *High Performance*.

Celebrating the centennial of the Statue of Liberty, 1986.

*This land is your land
This land is my land
From California
To the New York island,
From the redwood forest
To the Gulf Stream waters,
This land was made for you and me.*

—WOODY GUTHRIE

To become the instrument of a great idea is a privilege that history gives only occasionally.
—DR. MARTIN LUTHER KING, JR.

You know—we've had to imagine the war here, and we have imagined that it was being fought by aging men like ourselves. We had forgotten that wars were fought by babies. When I saw those freshly shaved faces, it was a shock. "My God, my God—" I said to myself, "it's the Children's Crusade."
—KURT VONNEGUT, *SLAUGHTERHOUSE-FIVE*, 1962

The American revolution is still going on—not because we ourselves are wise and good and helpful but because it embodies an idea that reaches everybody and will never lose its force.
—BRUCE CATTON, 20TH-CENTURY HISTORIAN

But where is what I started for so long ago? And why is is yet unfound?
—WALT WHITMAN, "FACING WEST FROM CALIFORNIA SHORES"

I believe that we are lost here in America, but I believe we shall be found....I think that the true discovery of America is before us. I think the true fulfillment of our spirit, of our people, of our mighty and immortal land, is yet to come.
—THOMAS WOLFE

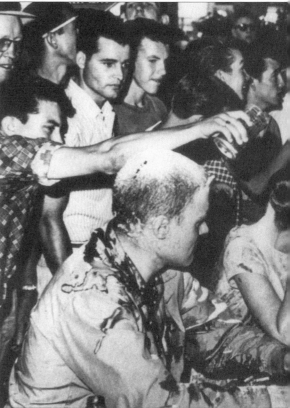

Clockwise from top left: Lucy and Ethel in I Love Lucy; *a Vietnamese baby being taken for treatment for burns (the baby died later); sit-in at a Woolworth's lunch counter in Jackson, Mississippi, 1963.*

Contents

Dr. Martin Luther King, Jr., preaches to his flock in Montgomery, Alabama.

PREFACE
About Democracy and Struggles

"What next?" says poor President Truman in 1946, as he faces the problems of a world shattered by war.

For more than a century, western Europe's nations had dominated the globe. Now, in 1945, they were exhausted. Two awful wars had been fought on their territory. Their peoples had suffered horribly. After World War II, it was as if there were a vacuum. We filled the vacuum. We had become the world's most powerful nation.

Our economy had been changed, and strengthened, by the war. We had acted quickly and with imagination. Using something we called *know-how*, we had built

Dominate means to command or be master of.

A **vacuum** is an empty space.

World War II began in 1939 when Germany invaded Poland. The United States entered the war after December 7, 1941, when Japan attacked Pearl Harbor. The war came to an end when Japan surrendered in August of 1945.

All of America's citizens threw themselves into winning the war. But when the war was won, many found there was another battle for freedom still to be waged at home.

9

In 1945, most schools and employers, and even the government, discriminated against women and people of color. Some businesses and public organizations were for white Protestants only. Immigration laws didn't treat people equally. Were those things fair? Have they changed? Are more changes needed? (Those are some of the questions this book raises. You can try to come up with answers.)

See Book 5 of A History of US *to read about Elizabeth Cady Stanton.*

tanks, ships, airplanes, and bombs better and faster than anyone thought possible. Women, blacks, and others—who were not always treated according to America's creed of fairness—worked as hard as anyone else. Many fought and died for their nation. Then, when the black soldiers came home, they were often not allowed to vote. Women workers were paid less than men for doing the same job. Was that fair?

Those citizens began to demand equal rights, which was their right as Americans. Anyone who read the Declaration of Independence knew *all men are created equal.* Elizabeth Cady Stanton had changed that to *all men and women are created equal.* Did *all* really mean all? Did it mean all people of

Building America:

The Liberty Bell is old, and it has a bad crack. It's been two centuries since it last rang. And yet, in our minds, it sounds loud and clear, because it stands for a set of ideas —political ideas—and they are America's gift to the world.

We are a nation built on ideas (instead of on a sameness of birth and background). One of our national ideas is that what you believe is no business of the president or the government. The Founders called that *freedom of conscience,* or *freedom of religion,* or *separation of church and state.*

We have another great idea. It is the idea that *we, the people,* can be responsible for ourselves. We can run our own government. We can pick our own leaders. That is called *democracy.*

Democracy had been tried before our constitution was written. More than 2,000 years earlier, a small Greek city-state named Athens tried democracy, and it worked marvelously well. Most Athenians were prosperous and happy; their sculpture, plays, and writings have rarely been surpassed. But Athenian democracy had flaws. Slaves did the hard work, women did not vote, and there was no protection for minorities when the voting majority made a poor decision. (The Athenians made a bad mistake when they voted to sentence the philosopher Socrates to death because he didn't believe in their democratic ideas.)

We improved on the Greek idea of democracy. We worried about protecting individuals and minorities from what is called "the tyranny of the majority." But we didn't establish

An Idea-Centered Nation

a perfect democracy. Our constitution begins with the words "We the people," but we didn't mean *all* the people. Like the Greeks, we allowed slavery, and women couldn't vote.

Still, ours was the best constitution any nation had ever written. Our Founders—who were thinking men —had read widely; they used the best ideas they could find from history in planning our nation. They studied the republican government of ancient Rome, they studied England's Magna Carta and its Glorious Revolution, and they studied the Iroquois confederacy. They read the words of writers on government, especially England's John Locke. They understood that a fair government is a process. It doesn't happen all at once. You have to work at it, and innovate, and adapt.

That process began right away, when some of the new citizens of the new nation demanded a bill of rights. So James Madison wrote ten amendments to the Constitution—called the *Bill of Rights*—that guaranteed rights such as freedom of speech and of religion and of the press. That was very unusual in the 18th century (it still is).

It may seem surprising to you that freedom and self-government are unusual. But they are. After our constitution went into effect, in 1789, other nations began looking at America to see if democracy would work. A Frenchman named Alexis de Tocqueville (duh-TOKE-vil) came to this country to see for himself. He said America was a laboratory for democracy. Tocqueville said that what was happening here was "interesting not only to the United States,

but to the whole world. It concerns not a nation but mankind."

Mankind was watching and taking notes. Soon some other nations became democratic. But there was something that was spoiling things in the United States; it was like a worm in a good apple. No, it was worse than that; it was clearly evil— even though it was common practice in many places. Jefferson called it a "cruel war against human nature itself." Yet he and his friends had not done away with it. It was slavery.

Getting rid of slavery was hard. It involved property rights (slaves were property to some people); it hit slave owners in their pocketbooks. Finally, a civil war was fought to end slavery. It would have been better,

The Liberty Bell's home is in Philadelphia, where the Constitution was drafted and proclaimed the guiding law of the land.

and wiser, if we could have done it without a war. But slavery was wrong; we needed to get rid of it, and we did.

How about women? The 15th Amendment said that all citizens have the right to vote. Were women citizens? The 15th Amendment didn't say. The men who were running the country (and some women) didn't seem to think they were. So women had to protest, picket, and even go to jail until, in 1920, another amendment—the 19th—gave them the right to vote.

By this time, people began to no-

tice that democracies don't usually go to war with each other, and so we learned that it was important to encourage democracy elsewhere.

But some people didn't understand. They seemed to think that democracy just meant the absence of all controls. Have you heard anyone say "This is a free country, isn't it?" when they want to do something they shouldn't do?

Well, total freedom isn't what democracy is about. That is *anarchy*, and it leads to disaster. Democracy is responsible government. It has controls established by *we the people*.

Remember, creating a free, fair government is a process, and not an easy one. Improving that government is a process without end. One thing is clear: in a government of the people, the people have to pay attention. If citizens don't get involved in their government, they can lose their precious rights. In a democracy, if you want to change things, you have to be part of the process. As Jefferson said, "If a nation expects to be ignorant and free, in a state of civilization, it expects what never was and never will be."

We do a better job of taking part in our government than most people realize. In this book you will see some Americans risk everything, even their lives, to help make our democracy what it was meant to be: a government for *all the people*.

In 1947, this cartoon suggested, the communist vulture has replaced the stork and threatens to drop Baby Chaos among the European nations crippled by war if Dr. U.S. Congress doesn't step on the gas and come to their rescue.

Demons are devils or evil spirits.

During the first half of the 20th century we fought two horribly destructive world wars and an economic depression. What we didn't fight was segregation and unfairness at home.

Rivals are competitors.

every color and description? Most Americans thought so.

But some people said those words in the Declaration without really listening to them. The United States government often did the same thing. Our nation wasn't guaranteeing basic human rights to all its citizens. Habit and selfishness were standing in the way of fairness—and no one did much about it. That was going to change. It would take a struggle to overcome the demons of bigotry—a struggle that continues today. You will read about it in this book.

You'll read about another struggle, too: America *vs.* the Soviet Union (also known as the U.S.S.R.—the Union of Soviet Socialist Republics—or Soviet Russia, or, often, just Russia). The Soviet Union was the second most powerful nation in the world. We had been allies and friends during the war. But there was something about the Soviet Union that made us nervous. It was a communist nation with a totalitarian government.

Totalitarianism is a political idea; *communism* is an economic idea. In a totalitarian government, the leaders tell people what they can do and say, and punish them if they do otherwise. Totalitarianism is the opposite of free government.

Communism is a method for controlling work and distributing a nation's farm produce, manufactured goods, and services. A communist government takes charge of and owns its nation's businesses.

Under our system, *capitalism*, businesses are mostly owned by individuals or corporations. The government's job is to pass and enforce laws that ensure business fairness and safety.

Totalitarianism and democracy are enemies. Communism and capitalism are rivals. After two world wars, we were fearful of rivals. Soviet Russia wasn't the only communist dictatorship. China and Cuba were two others. Could we all live together on the same planet? Or would there be a World War III? It is with those worries that this book begins. It is 1945, and we are about to begin a war of nerves with the communist nations.

1 The Making of a President

Harry Truman aged about ten. He got his first pair of glasses at age six—they cost $10, a lot of money in 1890.

"Mr. President," a boy asked Harry S. Truman, "were you popular when you were a boy?"

"No," said Truman. "I was never popular. The popular boys were the ones who were good at games and had big, tight fists. I was never like that. Without my glasses I was blind as a bat, and to tell the truth, I was kind of a sissy. If there was any danger of getting into a fight, I always ran."

The boy, and his classmates, applauded. Maybe some of them had run from fights and understood that it takes some bravery to admit it. Maybe they wondered about the popular boys in Harry Truman's class. He had become president of the United States; what had happened to his schoolmates? And how exactly did he get to be president? And what was it like running the world's most powerful nation? Well, Truman would answer as many of their questions as he could.

Vice President Harry S. Truman became president near the end of World War II, after President Franklin Roosevelt died. It was, as he said, an accidental presidency. He had been a senator—a quiet, hardworking senator—who seemed an ordinary, likable man. Then, to his surprise, Roosevelt asked him to be vice

After he retired, Harry Truman spent a lot of time sharing his experiences with children. He especially liked to tell them stories from American history. "I'm mostly interested in the children. The old folks...they're too set in their ways and too stubborn to learn anything new, but I want the children to know what we've got here in this country and how we got it, and then if they want to go ahead and change it, why, that's up to them."

Senator Truman at home in Washington in 1942. The wartime Truman Committee investigated companies producing weapons and other supplies. Truman uncovered many abuses and saved the government millions of dollars.

Mr. Truman goes to Washington (in 1935). "I am hoping to make a reputation as a senator," he wrote his wife, Bess, back in Missouri. "But you'll have to put up with a lot if I do because I won't sell influence."

He held to the old guidelines: work hard, do your best, speak the truth, assume no airs, trust in God, have no fear. Yet he was not and had never been a simple, ordinary man. The homely attributes, the Missouri wit, the warmth of his friendship, the genuineness of Harry Truman, however appealing, were outweighed by the larger qualities that made him a figure of world stature, both a great and good man, and a great American president.

—David McCullough, *Truman*

president. He was just getting settled in that job when, suddenly, he was president. He felt, he said, as if a bull had fallen on top of him.

When most Americans looked at President Harry Truman they sighed. He certainly was ordinary: more like a next-door neighbor than a president. He refused to even try to be sophisticated. Why, except for a year in France as an army captain in World War I, he'd hardly been anywhere. He'd been a farmer, a bank clerk, a shopkeeper, and a county administrator—all in Missouri. When he arrived in Washington, at age 50, you could almost see the rough edges. Sometimes he lost his temper and didn't think much about what he was saying. But he was never mean, or dishonest.

In fact, his honesty was legendary. When he wrote letters home to his mother and sister, as he did almost every day, he paid for the stamps himself. The *franking privilege*—which allows senators and presidents to send their mail free—was meant for government business, he said. He never used it for his personal letters. He lived modestly on his salary, and he didn't use his position to earn extra money. When a Republican who was a political rival left his briefcase at the

In 1905, Harry (right) had a good job in a Kansas City bank. Then his father's farm failed, and Harry had to give up the bank to help run the farm that belonged to his grandmother (sitting, with Harry's mother, outside the Young family farmhouse).

White House, some of Truman's Democratic aides wanted to go through it and see what it held. President Truman was horrified. He would not do a sneaky thing like that.

But when he was president, some people made jokes about him and acted as if he were a hayseed, although a few people noticed that he was very good at making decisions. Later, a historian wrote of him, "With more fateful decisions than almost any president in our time, he made the fewest mistakes." A senator said he was usually wrong about all the little things, but right about all the big ones.

Harry Truman was president during clamorous times. An army of men was returning from military to civilian life; they needed jobs and homes. People were moving from farms to cities faster than ever before. Could those cities become good places for everyone living in them? Europe and Japan were devastated. How would they rebuild? People of color were being treated unfairly. Would that continue? How would we change from making tanks and bombs to making dishwashers and automobiles? And what about Russia? The Russian leader, Joseph Stalin, had made promises he wasn't keeping. Truman the president had to answer those and many other questions. He said that knowing history helped him do the job.

Harry Truman could have been a history teacher; he knew a whole lot about the subject. His interest began when he was a boy and his father read a book to him about the ancient Greeks and Romans. He found he loved stories about people, especially real people. So, as soon as he could read himself, he started on biographies. Andrew Jackson became a special hero of his. Jackson was the kind of man Harry wanted to be: a man of action who represented the common people. A man who was independent, free-thinking, and not at all stuck up.

Truman was born on a farm in Jackson County, Missouri (which was named after Andrew Jackson); most

Bess Wallace aged 16. Her mother didn't think Harry was good enough for Bess; the first time he proposed she turned him down.

When he was a boy, Harry Truman read a book by an ancient Greek author named Plutarch. The book is called *Lives*. Plutarch wrote about people in pairs, contrasting Greek and Roman personalities. His book is lively and full of interesting dialogue and stories of historical events. Truman read it again and again throughout his life. You might like it too.

Harry Truman rides the cultivator over a field of young corn. The Young farm was over 600 acres, one of the biggest in the county, and Harry had to work very hard. He took it well, but he didn't enjoy milking cows. He liked the hogs best, and gave them pet names—one was called Carry Nation (who was she?).

Grandfather Solomon Young in his seventies, when Harry was a little boy. He was "quite a man, a great big man," Harry remembered, and in the summer took Harry riding all over the countryside in a high-wheeled cart.

In January 1946, the first General Assembly of the United Nations meets in London, England. Secretary of State James F. Byrnes heads the U.S. delegation, which includes Eleanor Roosevelt, widow of the former president. In October, the General Assembly accepts a gift of $8.5 million from John D. Rockefeller, Jr., to pay for a U.N. headquarters site. That site is in New York City.

people in Jackson Country felt as he did about the seventh president. When Harry was six, the family moved to nearby Independence. There, Harry discovered the public library and started reading all kinds of books. He never stopped.

He was soon forming his own opinions, and he didn't always agree with those around him. Reading gave him information; it allowed him to think for himself. There was one president whom everyone in Truman's family hated. Really hated. They could hardly talk about him without getting angry. But the more Harry Truman read about that president, the more he admired him.

It was the Civil War president. It was Abraham Lincoln. People hated Abraham Lincoln? They certainly did. You see, Harry Truman was a boy at the end of the 19th century, when many men and women could remember the Civil War. They hadn't cooled down. Harry Truman's parents and grandparents remembered the war as if it had just happened.

Truman's grandparents, both sets of them, had come to Independence, Missouri, in the 1840s, during the early pioneer days, when Missouri was a border state—and a slave state. They came from Kentucky by steamboat, newly married, bringing slaves they got as wedding presents. They weren't unusual; most of their neighbors were slave owners too. They were decent people who worked hard and tried to live a good life. They didn't think slavery was wrong. (Do you think some things we do now will be judged harshly in the future? What things?)

One of Truman's grandfathers, Solomon Young, was a pioneer who led wagon trains and herds of cattle across the Overland Trail—to California and Oregon and Mormon Utah. Independence was called the jumping-off place; it was the last town before the wagon trains started on the trails west. Everything west of Missouri and east of Califor-

Harry's father and mother, Martha and John Truman, as newlyweds. They gave Harry the middle initial *S* when he was born— but it didn't stand for any name.

nia was known as "the Great American Desert." It took some courage to venture out into that desert. Each journey to California and back took Grandfather Young about a year. On one trip he bought most of the land that eventually became the city of Sacramento. It was a family tale—how, if he had kept it, they might have all been rich. But if Solomon Young's grandson Harry Truman had been rich—well, maybe he wouldn't have worked hard and become president of the United States.

Now back to the Civil War. The Kansas–Missouri region was one of the hottest and meanest regions before and during that war. It was in Kansas that the abolitionist zealot John Brown got out his hatchet and chopped some people to bits. And he wasn't the worst of the killers, not at all. There were some terrible things done—on both sides.

One morning in 1861, Truman's Grandmother Young was on her farm (her husband, Solomon, was away) when a band of Union raiders galloped into the yard, ordered her to cook a big meal for them, killed all her chickens and 400 hogs, set fire to the barns, and then rode off with the freshly butchered meat, 13 mules, 15 horses, and the family silver. While all this was going on, 11-year-old Martha hid under the kitchen table.

During the war, many well-known actors, singers, and comedians entertained the troops—including Vice President Truman, who played under the keen eye of movie star Lauren Bacall at Washington's National Press Canteen in 1945.

Two years later, Martha and the rest of the family were marched to a Yankee fort where they were kept prisoners. Their home—a white-pillared plantation house—was burned to the ground by Union soldiers. Are you surprised that Martha Young hated Yankees and President Lincoln?

Martha was Harry Truman's mother. She grew up to be a strong woman who played the piano well, had a good education, and said what she thought—which was a trait that she passed on to her son. (He was a good piano player, too.) Once, when she came to visit the White House, the only empty bed was Lincoln's famous one. Now, most White House guests feel very privileged if they can sleep in the very bed where Abraham Lincoln slept, but not Martha Truman. She said if that was the only bed, why, she'd just sleep on the floor. She was well known for her sense of humor, but this time her son knew she wasn't kidding. He found another bed for her.

John Brown was a violent zealot who claimed he was inspired by God to help the slaves. See Books 5 and 6 of *A History of US* for details.

A **trait** is a special characteristic, something that distinguishes you from others. Usually it has something to do with your personality—like plain speaking, or lying, or being optimistic.

17

2 A Major Leaguer

Even at college, said his wife, Jackie Robinson "walked straight, held his head up, and was proud not just of his color, but his people."

In 1945, we were a Jim Crow nation. It was nothing to be proud of, but that's the way it was. In the South, everything was segregated: schools, buses, restaurants, hotels, even phone booths. The rest of the country wasn't as blatant about it, but there was plenty of separation and prejudice.

In the U.S. armed services, blacks were allowed to die for their country—as long as they did it in segregated regiments.

And when it came to the national pastime—which is what baseball is called—there were the major leagues, the minor leagues, and there were the Negro leagues (for ballplayers of color).

Those who approved of Jim Crow segregation said that things were "separate but equal." They were separate all right. But they were rarely equal. And they certainly were not on the ballfield.

The major leaguers played in fine ballparks, traveled first class, and slept in decent hotels. The Negro leaguers? Well, they put up with a lot: shoddy conditions, no ballparks of their own (they rented what they could find), travel any way they could make it, and—usually—lower pay (except for the incredible Satchel Paige, who in 1942 managed to make more money than anyone in any league).

One thing the Negro leagues did have in abundance was talent. When black players played all-star games against white teams they usually won.

Just think about it, and you can see how insane the system was. All those good ballplayers and no one letting them play in the majors! There were plenty of whites who understood that; and there were plenty of whites without prejudice.

One of them was the general manager of the Brooklyn Dodgers. His name was Branch Rickey. Rickey decided he was going to change baseball. He was going to make it the national pastime for all Americans.

But he knew it wouldn't be easy. Fighting prejudice never is. Rickey was the right man for this job. He had founded baseball's system of farm teams back in the 1920s. That means he came up with the idea of taking over minor-league teams (which had been independently owned, just like the major-league clubs) and using them to develop ballplayers for the major leagues. Branch Rickey was used to scouting good players. He knew how to pick them. He was also a shrewd businessman. Black ballplayers (then) were a pool of inexpensive talent. They played an exciting, hustling kind of baseball. And they would bring a huge new black audience to the majors.

If Rickey was going to change baseball and some of the nation's attitudes by integrating the Brooklyn Dodgers, he knew he would have to find a ballplayer who was not only a great athlete, but, even more important, a great person. When he found Jack Roosevelt Robinson he found just the man he was looking for.

Jackie Robinson was a spectacular athlete. He had earned letters and trophies in four sports at the University of California at Los Angeles (UCLA). He was very smart and did well in school. And he had the strength to fight for his beliefs. As an officer in the army, Robinson refused to move when a bus driver asked him to sit in the back of the bus (where blacks were expected to sit). That got Jackie in trouble, but he wouldn't back down. He faced a court martial (a military court) for disobedience. But the young lieutenant had acted within his rights and the army dropped the charges against him. Some people thought him a troublemaker, but Branch Rickey was impressed. Here was a

In 1943 Bill Veeck (VEK) tried to buy the Philadelphia Phillies and sign up black players. An editorial in the *Sporting News* scolded him for the very idea. As long as Judge Kenesaw Mountain Landis was baseball commissioner there wasn't much chance of it. He was a bigot. Soon after he died, in 1944, Branch Rickey began looking for black players. The new commissioner, A. B. "Happy" Chandler, former governor of Kentucky, said, "If a black boy can make it on Okinawa and Guadalcanal, hell, he can make it in baseball."

When some Dodgers said they wouldn't play on the team with a black man, Rickey traded them away.

Branch Rickey signs Robinson up. "Baseball people are generally allergic to new ideas," said Rickey. "It took years to persuade them to put numbers on uniforms....It is the hardest thing in the world to get big-league baseball to change anything....But they will...eventually. They are bound to."

Mack Robinson, one of Jackie's older brothers, was a world-class sprinter who finished second to American track star Jesse Owens in the 1936 Berlin Olympics. (Jesse Owens is someone to read more about, in Book 9 of *A History of US* and elsewhere. His is quite a story.)

Late, late as it was, the arrival in the majors of Jack Roosevelt Robinson was an extraordinary moment in American history. For the first time, a black American was on America's most privileged version of a level field. He was there as an equal because of his skill, as those whites who preceded him had been and those blacks and whites who succeeded him would be. Merit will win, it was promised by baseball.

—A. BARTLETT GIAMATTI, *TAKE TIME FOR PARADISE*

man of courage, he believed.

Rickey asked Robinson to come to New York. He said he wanted to talk about a new Negro team. Then, in his office, Branch Rickey told Jackie the truth: he wanted him to break baseball's color line. Both men knew that the first black ballplayer in the major leagues wouldn't have it easy. Rickey told Robinson that if he wanted the job—no matter what happened to him—he had to promise not to fight back. He would have to take abuse and hold his tongue. At all times he would have to be a gentleman.

"Mr. Rickey, do you want a ballplayer who's afraid to fight back?"

"I want a player with guts enough not to fight back," said Rickey.

Robinson had never backed away from a fight. He knew that if someone insulted him it would be very difficult to do what Rickey asked: to "turn the other cheek." But he agreed; he gave his word. He was going to do something bigger than anything he'd done before; it was more important than his feelings. It was for his people and for all people.

The two men talked for three hours. Still, neither of them realized how much courage Jackie Robinson would actually need. He had tough times ahead of him. He was going to be spiked, spat on, sent death threats, hit with pitches, and called awful names. How would you have responded?

Branch Rickey began by sending Jackie Robinson to Brooklyn's leading farm team, the Montreal Royals. The Royals' manager,

Robinson was the first UCLA student ever to win letters in four different sports. He beat his own brother's national long-jump record and also won tournaments in tennis and golf.

Of his four college sports, Robinson liked baseball the least. But when he left the army in 1944, the Negro league Kansas City Monarchs offered him the job of shortstop, and he took it.

Clay Hopper, had grown up with prejudice. He had never had a black friend. He begged Branch Rickey not to make him coach Jackie Robinson. Rickey knew he was a good coach; he told him to do his job. By the end of the season Hopper had learned a lesson: most prejudice comes from ignorance. He told Robinson, "You're a real ballplayer and a gentleman. It's been wonderful having you on the team."

On April 15, 1947, Jackie Robinson, up from Montreal, batted in Brooklyn for the first time as a major leaguer. He was put out four times that day. He didn't do much better the rest of the week. Had Rickey made a mistake?

Then, when the Dodgers went to Philadelphia to play the Phillies, even Rickey was stunned by what happened. The Phillies' manager, Ben Chapman, spewed hate language and encouraged his players to do the same. "At no time in my life have I heard racial venom and dugout abuse to match the abuse that Ben sprayed on Robinson that night," said one of Branch Rickey's aides. "I could scarcely believe my ears," said Robinson.

Jackie Robinson took a deep breath and kept his word. The abuse wasn't all verbal. Runners were sliding and cutting him with their spikes, pitchers were throwing at his head. It was too much for his teammates—even for those who hadn't wanted a black player on the club. "You yellow-bellied cowards," yelled a Dodger player. "Why don't you pick on somebody who can answer back?"

"If you guys played as well as you talked, you'd win

Robinson gritted his teeth and stuck it out through all the abuse. "I'm not concerned with your liking or disliking me," he said. "All I ask is that you respect me as a human being."

Jackie Robinson steals home. Black baseball was different. "In our baseball," said Buck O'Neil, "you got on base if you walked, you stole second, you'd try to steal, they'd bunt you over to third and you actually scored runs without a hit."

I think sports...teach a guy humility. I can see a guy hit the ball out of the ballpark, or a grand slam home run to win a baseball game, and that same guy can come up tomorrow in that situation and miss the ball and lose the ball game. It can bring you up here but don't get too damn cocky because tomorrow it can bring you down there. See? But one thing about it though, you know there always will be a tomorrow.

—BUCK O'NEIL, FIRST BLACK COACH IN BIG-LEAGUE BASEBALL

21

> **I slid into him this one time and really cut him badly....I could see he was bleeding the same color blood as me. I just stood there and felt ashamed of myself, like a real jerk.**
>
> —RICHIE ASHBURN, STAR OUTFIELDER FOR THE PHILADELPHIA PHILLIES

some games!" hollered another Dodger.

Sometimes actions bring unexpected results. The poor sportsmanship of some other teams brought the Dodgers together. They were behind their new teammate now.

Soon Robinson was swinging—and connecting. And when it came to base running? Hardly anyone has ever done it the way Jackie Robinson did. He gave pitchers the jitters. And when he stole home? Well, have you ever seen anyone steal home? There isn't much in baseball that is more exciting. Robinson was a fantastic base stealer.

In his rookie season, Jackie Robinson finished first in the league in stolen bases and second in runs scored. He tied for the team lead in home runs. Dodger fans began cheering and cheering. The nation's most important sports paper, the *Sporting News* (which had said that Rickey was unwise to bring a black to the majors), named him rookie of

the year. In September, the Brooklyn Dodgers won the National League pennant. And, at the end of the season, Branch Rickey told his star, "Jackie, you're on your own now. You can be yourself." Robinson no longer had to keep quiet, and he didn't.

Jackie Robinson had won the affection and respect of his fellow ballplayers and of the nation. He was the first; he took the punishment, he made it easy for those who followed. Baseball was now the national pastime for all the people.

In his first season Jackie helped set new attendance records at the ballparks. "A life is not important," he said, "except in the impact it has on other lives."

3 A (Very Short) History of Russia

Vladimir Lenin led the second Russian Revolution in 1917. "It is true that liberty is precious," he said. "So precious that it must be rationed." What did he mean?

In order to understand American history in the 20th century, you need to know some Russian history. Does that sound strange? Well, things were happening in Russia that would decide much that happened in the United States. Partly it was because we were obsessed with Russia, which means we couldn't get that country out of our minds. Partly it was because there were real dangers to the world from communist Russia's dictatorship.

After World War II we were determined to be mightier than the Soviet Union. Because of that, we spent vast sums of money on our military forces. We built huge stockpiles of expensive weapons—more than enough to blow up the world. We persecuted some of our own citizens because of fear of communist ideas. Sometimes we even seemed to lose faith in our way of life because we mistakenly thought Russian communism was more powerful.

Now for that Russian history. In 1917, during World War I, Russia had a revolution. For centuries, Russia had been controlled by tsars—who were like emperors. The word *tsar* (ZAR)

A ***dictator*** is a tyrant—a ruler who makes the people do what he wants (it is often not what they want).

The official name for Lenin's nation was the Union of Soviet Socialist Republics, or the U.S.S.R. Russia was the largest of a group of states, or republics. None were free, independent republics. The union lasted until 1991.

Exile means to banish someone from his or her home or country.
Oust means to throw out.

What the Russian peasant wanted was his own land. The communists said that peasants and workers would now be the owners and masters.

comes from *caesar*, which was the title of ancient Rome's great leaders. Many of Russia's tsars were selfish tyrants with absolute control over their people. The Russian people wanted something better: they wanted the things that all people want—peace, opportunities, and freedom. Alexander Kerensky led a revolution in 1917. When another revolutionary, named Vladimir Ilyich Lenin, heard that news in Switzerland, where he had been exiled by the tsar, he headed home to Russia. Lenin—who was the head of a radical political party called the Bolsheviks—ousted Kerensky and the more moderate rebels in a second revolution a few months after the first one. Lenin took over the leadership of

Kerensky was a tall, handsome lawyer with the gift of the gab. But it didn't save him from exile when Lenin took over the government.

Russia. He formed a communist government. It was an experiment. Communism had never been tried in a whole nation before. Lenin had to use force to make it work. He soon created a vicious, unfree, totalitarian government. When Lenin died, Joseph Stalin took over. He was worse than Lenin, and worse than any of the tsars. He killed millions of his own people. Russians who protested were murdered, or sent to prison camps in Siberia. Most never came home again. Meanwhile, Stalin and his followers were telling the rest of the world that the Soviet Union was turning into a wonderful, perfect society. It was hard for outsiders to find out the truth; many people believed the experiment was working.

Communism, to those who hadn't tried it, seemed like a fine economic plan. Most of the ideas for modern communism came from a 19th-century thinker named Karl Marx. Marx wanted to make the world better. He looked at capitalism and saw that, without regulation, wealth soon piled up in a few hands and left many people miserable. There was something even more disturbing: money power usually led to political power. So the poor had double troubles. They had no money and no political power. Marx said capitalism was doomed. And, during the worldwide Depression of the 1930s, it seemed as if he was right.

Marx's economic system was based on sharing. People were to work hard and give their products to the government, which would distribute things fairly to everyone as they were needed. It sounded pretty good. Unfortunately, it just didn't work. Things didn't happen

The tsar of Russia and his whole family were murdered; otherwise the Revolution was not especially bloody. But riots often brought the soldiers out, as here in Petrograd (St. Petersburg) in July 1917, before Lenin took power.

the way Karl Marx had predicted.

Russia and China were the largest nations that attempted communism. In those countries, work and pay were decided by the government. There was no economic or political freedom under communism. People usually don't work hard when they can't own the results of their work,

Karl Marx was German, but he was exiled from his own country for his political beliefs. He is buried in England. If you ever visit London you can see his grave in Highgate Cemetery.

25

FRESH AS A DAISY

on a sultry
Arkansas night

thanks to your

Electric Room Air Conditioner!

and government distribution is rarely fair. Besides, an enormous country like the Soviet Union was simply too big for one central government to run everything properly. The Russian government became terribly inefficient and wasteful. Perhaps communism didn't get a good test, as some said, but, mostly, the experts who had hoped for great things from Karl Marx's ideas were disappointed.

There was something else that surprised a lot of experts: capitalism wasn't doomed. If Karl Marx could have risen from his 19th-century grave he would have been astonished to find that in the United States, in the second half of the 20th century, capitalism helped a great many people pursue happiness. Free markets brought automobiles, washing machines, nice clothes, and TV sets to many Americans. But none of that was clear in 1945. We didn't understand ourselves, and we certainly didn't understand the Soviet Union. Some people thought Russia was the hope of the future. Some were terrified of communism without really knowing why. Others feared that communists were about to take over the United States. It was very confusing.

Many in the U.S. of the early 1950s feared that Stalin, the world's strongman, had the upper hand everywhere. This cartoon is titled *But What Part Shall the Meek Inherit?*

By the 1950s America was the world's wonderland of gadgets and all modern consumer goods. On average, the national standard of living was the highest on earth (although large numbers of Americans lived in poverty, especially in the South and in cities).

What if we'd spent all our military money on things to make our cities, schools, and towns safer and more prosperous? Would Russia have attacked us, as we feared? Would Russia have taken over in western Europe and the Middle East? What do you think? (No one knows the answers to those questions, but it is fun to think about them.) Historians have a big advantage. It is called **hindsight.** *We know how things come out. Some 40 years after World War II, Russian communism collapsed because the system proved unworkable.*

4 A Curtain of Iron

Churchill (right) told President Truman that his speech in Fulton, Missouri, would be about "the necessity for full military collaboration between Great Britain and the U.S. in order to preserve peace in the world."

England's great wartime leader, Winston Churchill, had something to say, but no one was listening. So, in 1946, when President Truman asked the former prime minister to speak at tiny Westminster College in Fulton, Missouri, Churchill didn't hesitate. He said yes.

Churchill wanted to talk about Russian communism. Many people did not know what to think about Stalin and Soviet Russia. During World War II (which ended in 1945), Russia was the ally of England and the United States. No people fought harder against the Nazis than the Russians. No nation suffered war losses as enormous as Russia's. When the war ended, everyone hoped for friendship between the new superpowers: Russia and America. Around the world, many people believed that Russian communism was an acceptable form of government.

Winston Churchill thought differently. Churchill had warned of Adolf Hitler and Nazism long before most Britons or Americans took them seriously. Once again, he wanted to tell the world of a dangerous dictator and an ominous form of government. "A shadow has fallen upon the scenes so lately lighted by the Allied victory," he said at that small Missouri college. The shadow he was talking about was vicious totalitarian rule. "From Stettin in the Baltic to Trieste in the Adriatic an *iron curtain* has descended across the Continent," Churchill continued.

I do not believe that Soviet Russia desires war. What they desire is the fruits of war and the indefinite expansion of their power and doctrines....I am convinced that there is nothing they admire so much as strength and there is nothing for which they have less respect than for weakness, especially military weakness.

—WINSTON CHURCHILL

Ominous means threatening.

A *totalitarian* government has total control over its citizens' lives.

The continent Churchill was referring to was Europe.

The curtain of iron was blocking out truth and freedom. Nations behind that curtain were prisoners of Russia.

When World War II ended, the armies of the winning Allied powers—the U.S., the U.S.S.R., and Great Britain—moved through Europe, freeing the nations that had been conquered by Hitler's Nazis. The Allies promised to help the liberated nations. They promised to help them hold open elections and form free governments. After that, the Allied armies were supposed to leave (which was what we did).

But Russia wouldn't go. Soviet armies stayed in control in Poland, Romania, Bulgaria, Czechoslovakia, Hungary, Yugoslavia, Latvia, Lithuania, Estonia, and East Germany. There were no free elections there. Elsewhere—in nations like Italy and France—the communist parties were growing strong. Stalin bragged that the whole world would eventually go over to communism.

For a short time in 1956, Hungary revolted against Soviet rule; these Hungarians burned Stalin's portrait in the streets.

But most people didn't stay behind the iron curtain willingly. At every Soviet border, armed guards kept peoples captive. Iron curtains would soon extend over several Asian countries. Some east European countries, like Hungary and Yugoslavia, attempted to rebel and become independent. The Hungarians were crushed and their leaders killed. The president of Yugoslavia, Marshal Tito (TEE-toe), was as crafty as Stalin himself, and he managed to keep the Soviet Union at arm's length. But even Yugoslavia was not really a free country. It had only one political party, and that was communist.

President Truman decided the United States would come to the aid of any nation endangered by communism. We would not let Soviet Russia expand further. We began by sending $400 million in emergency aid to Greece and Turkey. That program of assistance was called the Truman Doctrine. It was the beginning of a *cold war* against Russia. The Cold War lasted more than 40 years.

> [Communism] is based upon the will of a minority forcibly imposed upon the majority. It relies upon terror and oppression, a controlled press and radio, fixed elections, and the suppression of personal freedoms. I believe that it must be the policy of the United States to support free peoples who are resisting attempted subjugation by armed minorities or by outside pressure.
>
> —HARRY TRUMAN, ANNOUNCING THE TRUMAN DOCTRINE TO CONGRESS

After the war, Germany was split in two. East Germany stayed under Soviet control, and West Germany got a free, democratic government. The old capital of Germany, Berlin, was also divided between east and west (see map inset opposite). In 1961, the Russians built a concrete wall in the middle of Berlin and topped it with barbed wire to keep people from running away to freedom.

5 The Marshall Plan

President Truman (center) with Secretary of State George Marshall (right) and Undersecretary of State Dean Acheson. "Marshall is a tower of strength and common sense," said Truman.

Two signs sat on President Truman's desk. The first sign quoted a man from Truman's home state of Missouri. It said, *Always do right. This will gratify some people & astonish the rest.* They were the words of Mark Twain.

The second sign said THE BUCK STOPS HERE.

Which means: the president has the final word and can't blame anyone else for his decisions.

Harry Truman had some big decisions to make. Those decisions would profoundly affect Americans and people around the world. In one of the most important of his decisions, he persuaded the American people to act generously to the defeated nations. What Harry Truman had in mind had never been done before in the history of the world.

Truman knew how defeated people feel after a war. He knew that his Confederate ancestors carried hate in their

A Soviet cartoon portrayed Truman as an imperialistic Uncle Sam towed along by the European nations hungry for dollars—they would never get them, said the cartoon.

hearts all their lives. He knew that Germany's anger after World War I had helped bring about a second world war.

So he supported a plan that would send billions of dollars in aid and assistance to our allies and to our former enemies. It was called the Marshall Plan, but it reflected President Truman's thoughts. After a terrible war, he was asking the winning nation to help everyone recover—including the losers. The president said:

> *You can't be vindictive after a war. You have to be generous. You have to help people get back on their feet....People were starving, and they were cold because there wasn't enough coal, and tuberculosis was breaking out. There had been food riots in France and Italy....We were in a position to keep people from starving and help them preserve their freedom and build up their countries, and that's what we did.*

Marshall Plan aid was offered to all of Europe's nations—including the Soviet Union and those countries under Soviet control. The Soviet nations refused the aid. Sixteen nations accepted with enthusiasm. It was very expensive. It was very unselfish. The plan encouraged Europeans to use American aid and add their own brains and know-how. It worked. Prosperity began returning to the free nations of Europe. It also helped us. Those newly prosperous European nations now had money to buy American goods. And they did.

Marshall Plan aid: a ship unloads U.S.–made tractors in France. The plan called for spending $12.5 billion in 16 countries over four years.

Someone who is ***vindictive*** wants to have revenge.

Marshall Plan aid was very concrete (that's a pun). U.S. money rebuilt steel mills in Belgium, ceramics factories in France, railroads in Germany, and bridges and buildings in a whole lot of places.

31

The Diet is the name of Japan's congress. What is the name of Great Britain's congress?

Discrimination means choosing for unfair reasons. (The word has other meanings. Look it up.)

"We cannot wait another decade or another generation to remedy these ills," said Truman. "We must work, as never before, to cure them now."

In East Asia, General Douglas MacArthur was sent to defeated Japan as head of an occupation army determined to rid Japan of its war leaders and help bring democracy, freedom, and prosperity to that nation. The Japanese wrote a new constitution; it made Japan a democracy. Land was redistributed so that more people could have it. (Instead of a few huge landowners and many poor farmers, there was now a better balance. Soon there would be great prosperity.) Women were allowed to vote (39 were elected to the Diet). Secret political societies were prohibited. And religious discrimination was ended. The United States poured aid into Japan—food, clothing, medicines, and other supplies. Ancient temples and museums were restored. We were very generous. No nation had ever done that kind of thing for a defeated foe.

Another Truman plan, called Point Four, gave aid to developing nations. Developing nations (another name for them is the Third World) are countries that are less wealthy and less modern than we and the other industrial nations—many of those developing nations are in Africa, Asia, and Latin America.

Did everyone approve of these generous policies? Not at all. Some people in Europe and Asia said they didn't want to take aid from America. They thought we wanted something in return. Some people in Congress yelled about all the money it was costing. "Why should we help others?" they asked. "Why should we help our former enemies?" they screamed.

Their screams were nothing compared to those heard when President Truman decided to do something to help people in the United States. He decided it was time to do something about civil rights for all citizens. He decided to do something about lynchings and segregation. The army, navy, and air force were all segregated. Blacks and whites served in separate units. Blacks got the worst jobs. That wasn't fair. Like other Americans, they were willing to fight for their country. Why should they be treated differently?

In Mississippi, when some black soldiers returned home, they were dumped from army trucks and then beaten. In Georgia, a black man was shot and killed because he had voted. When Truman heard of those outrages he was horrified. The president had been brought up on Confederate ideas, but he was also taught to know right from wrong. Maybe Mark Twain's words on his

Truman's 1947 speech denouncing racial discrimination and pledging to fight it was the first ever made by a president to the NAACP (the National Association for the Advancement of Colored People).

desk helped inspire him. He sent proposals to Congress to stop lynchings, to outlaw the poll tax that kept some people (mostly blacks) from voting, and to end segregation in the armed services. He created a commission on civil rights.

*A **lynching** is an outside-the-law execution.*

Remember the villains of prejudice and hate? People infected with those viruses began to howl. A Florida county commission said the president's program was "obnoxious, repugnant, odious, detestable, loathsome, repulsive, revolting and humiliating." A Mississippi congressman said Truman had "run a political dagger into our backs and now he is trying to drink our blood." Read on, and you'll see what happened next.

Obnoxious, repugnant, odious, detestable, loathsome, repulsive, revolting, humiliating? Great words—what do they mean?

33

6 A "Lost" Election

Thomas Dewey and his wife, Frances. "His mechanical smile," said a reporter, "and bland refusal to deal with issues, have got under everybody's skin."

Who was the Civil War president? What party did he belong to?

Democrats could count on winning in the South. No southern state had voted for a Republican for president since before the Civil War. The South was known as "the solid South." It was solidly Democratic.

Now, because of Truman's civil rights proposals, many southern politicians were furious with their party. They weren't quite ready to turn Republican, but they were certainly against Harry. So some formed another party. It was called the Dixiecrat Party.

Other Democrats were unhappy with the president for different reasons. Some thought Truman was too hard on communism. They wanted the United States to try to get along with Joseph Stalin and the Soviet-controlled countries. Some wanted more domestic reforms. Those people formed another party. It was a new Progressive Party.

When a candidate splits his own party in three—well, he is in trouble. In 1948, Harry Truman and the Democrats were in trouble. Besides, the Democratic Party had been in power since 1932, so most people said they were ready for a change.

Truman was nice enough. But after that giant of the war years, Franklin Roosevelt, Harry Truman seemed almost embarrassing. Sometimes he just popped off and said whatever was on his mind. He wasn't dignified. He wasn't meant to be president, some people said.

So everyone knew that Harry Truman didn't have a chance to get elected in 1948. Some Democrats tried to dump him. They wanted someone else as their candidate. But Harry S. Truman was stubborn. He was head of the party and he was going to run for election.

The Republicans chose Thomas E. Dewey as their candidate for president. Dewey was governor of New York. He was much younger than Truman, but he acted old and wise. He had a trim, compact build, dark hair, and a small dark mustache. He was dignified. He didn't say much. He didn't campaign hard. He just began to act as if he were president, because everyone knew he would be soon. (Congresswoman Clare Boothe Luce said that Harry Truman was a "gone goose.")

Franklin Roosevelt had used radio to talk to the American people. Truman wasn't a good speaker on the radio. But he was pretty good in person, especially when he spoke without a prepared speech and just said what he thought. So Harry Truman got on a train and began his campaign. (In those days most people traveled long distances by train. There weren't many big highways, and air travel was still a novelty, and expensive.)

The president's train had bedrooms, a dining room, a car for newspaper reporters, a car for presidential aides, office space, and a wood-paneled sitting room for the president and his family—16 cars in all. The train crossed the nation—twice. When it pulled into a city, or town, or hamlet, the president stood on the back platform and spoke to anyone who came to the railroad station to hear him. Lots of people came. Wouldn't you go to hear a president?

Sometimes Harry Truman gave his first speech before six in the morning. He was a farm boy, and used to getting up early. He gave speeches all day long—10 or 15 a day. Sometimes he got off the train for an outdoor rally. Usually there were flags and bunting and local politicians to share the platform. Sometimes

The presidential train was named the *Ferdinand Magellan*. Who was Magellan?

If you can't stand the heat, get out of the kitchen.
—FAVORITE TRUMAN SAYING

Harry Truman had a hard time with Congress. That body approved his foreign-aid plans but turned down many of his domestic (home) proposals. Some of those proposals—for civil rights, national health insurance, and urban planning—were farsighted. But Congress did pass some important bills. One was called the G.I. Bill of Rights. It gave military veterans a chance for a free college education. It educated a generation of Americans (mostly men), and that helped create a broader and stronger middle class.

Bess—or "Boss"?

Truman liked to introduce his wife, Bess, and his daughter, Margaret, to those who appeared at the whistle stops. "Would you like to meet the Boss?" he'd say before Bess appeared. "He's the president," one editor wrote, "[yet] he's just an ordinary family man, proud of his wife and daughter. He has something in common with many who hear him." (Bess told her husband that if he called her "the Boss" one more time, she'd get off the train.) But the editor was right: Harry was a family man and *very* devoted to his wife and daughter.

Bess Truman (waving) said that the essential attributes for a first lady were good health and a well-developed sense of humor.

Black people gathered in huge numbers wherever Truman campaigned; these voters are in Harlem, New York. For the first time, black delegates were present in force for Truman's nomination at the Democratic national convention.

The Hotel Roosevelt was named for Theodore Roosevelt (a Republican), not for Democrat FDR.

Truman was a good mimic. After the election, he liked to imitate H. V. Kaltenborn saying that Dewey would win.

high-school bands played and marched, and the president gave a luncheon or dinner speech in a big city hall. It was exhausting to everyone except Harry Truman, who seemed to get more energetic and feisty as the campaign continued. His speeches were fighting speeches. He lashed out at the Republican Congress (which wasn't passing the laws he wanted), and he attacked those who asked for special government favors: he called them "power lobbies" and "high hats." People cheered his spirit (even if they didn't seem impressed with him otherwise). "Give 'em hell, Harry," they said.

Tom Dewey had a train too. But he didn't get up early, and he didn't give many speeches. He didn't need to. It was clear that he was going to win. *Everyone said so.*

Newsweek magazine asked 50 leading journalists—people whose business it is to know politics—who would win. *All 50 said that Truman would lose.* One of Truman's aides bought *Newsweek* and read it on the train. Not even one reporter gave the president a chance to win. The aide tried to hide the magazine, but Truman spotted it and read the article. "Don't worry," he said, "I know every one of those 50 fellows, and not one of them has enough sense to pound sand into a rat hole."

The sensible *New York Times* conducted a survey. It sent reporters around the country for a whole month. The reporters concluded that 29 states would go to Dewey, 11 to Truman, and four to the Dixiecrats. The others were undecided.

Every leading poll showed a Dewey landslide.

On November 2, 1948, the American people voted.

That evening, Dewey's supporters crowded into the ballroom of New York's Hotel Roosevelt. They were there to celebrate. Men wore black tuxedos and women wore evening gowns. Each Republican woman was presented with an orchid as a victory corsage. Waiters carried trays of elegant food.

In Washington, the Democrats hadn't even rented their usual hotel ballroom. They were short of money and there was no point in wasting it—they had nothing to celebrate.

Newspaper reporters wrote articles congratulating the new president on his victory—that way they could go to bed as soon as the returns came in. At the *Chicago Tribune,* the morning's headline announcing Dewey's victory was set in type.

As night arrived, the counting began. (In this time before computers, vote counting was slower than it is today.) Maybe it was habit, but many people stayed up to listen to their radios. They expected a quick decision. Election results were being broadcast on television, for the first time. But most people didn't have television sets. Harry Truman didn't have one. He was staying at a small hotel. He ate a ham sandwich, drank a glass of buttermilk, and went to bed early. When it was announced that he had won in Massachusetts, one of his Secret Service men woke him with the news. "Stop worrying," said Truman, and went back to sleep.

At midnight he woke up, turned on the radio, and listened as a deep-voiced radio commentator named H. V. Kaltenborn announced that although Truman was a million votes ahead, they were just early votes: Dewey was sure to win.

At 1:30 A.M. the Republican National Chairman stood on a chair in the Roosevelt Hotel ballroom and said that it looked as if Dewey would win in New York State and would soon be president of the United States. The guests cheered.

At 4 A.M. the Secret Service agents received a call from Democratic headquarters. Illinois had been put in Truman's win column. They couldn't resist waking their boss. "That's it," he said. "Now, let's get back to sleep."

At dawn, Truman got up. Deep-voiced Kaltenborn was still on the radio. Now he was saying that it was a very close election—but Dewey would win. By mid-morning it was clear: all the experts were wrong! Truman was no accidental president. He had won the job on his own.

Left, President Truman relished the chance to laugh at the eagerness of the press to finish him off in advance. The crowd that gathered outside the White House (below) after his election was the biggest in history.

7 Spies

Hollywood producers, terrified of being accused of sympathy to communism, tried to prove their loyalty with strings of movies "exposing" red "conspiracies."

The times were prosperous, but not content. There was fear in the air.

Some Americans were afraid there might be a communist revolution in the United States. They believed that our nation was filled with communists.

Some thought that President Roosevelt's New Deal laws were inspired by communists. That legislation had changed America with strong child labor regulations, minimum wage standards, Social Security, and new taxes. All that had put some limits on capitalism. More people shared the wealth. The gap between rich and poor had been narrowed. There was a newly prosperous middle class (as there was when the nation was founded). That strong new middle class was challenging the old guard and its ways. Now Harry Truman wanted to change society even more, with his civil rights ideas and with a program of liberal reform called the Fair Deal. "Suppose he gets his national health insurance—who will pay for it?" some people asked. "Those who have the most money will pay most of the bills," they said. To many, it sounded like communism.

Then communist spies were discovered in the United States. They had stolen atom-bomb secrets and sold them to Russia. As if that weren't bad enough, shocking news came from England—some top British intelligence officials turned out to be Soviet spies. And that

Ethel and Julius Rosenberg (above) were convicted in 1951 of selling atomic secrets to the Soviet government. Their case set off a wave of anti-communist hysteria (left), and they were executed by electrocution.

[South Korea's Syngman Rhee] was one of the early postwar, anti-Communist dictators, with an instinctive tendency to arrest almost anyone who did not agree with him. Only by comparison with his counterpart in the North, Kim Il Sung, did he gain: Sung not only arrested his enemies; he frequently had them summarily executed....[Sung] held the Order of Lenin, awarded by Stalin himself....At first he was a popular figure, for it was widely known that he had devoted his life to fighting the hated Japanese. That popularity would diminish as the harshness and cruelty of his rule became apparent.

—DAVID HALBERSTAM, *THE FIFTIES*

wasn't all: in a case that filled newspaper headlines day after day, a former State Department adviser and president of an international peace organization—a man named Alger Hiss, whom everyone trusted—was convicted of lying about his involvement with an admitted communist. A young congressman, Richard Nixon, captured the attention of the whole nation with his hard questioning of Alger Hiss. When Hiss was found guilty of *perjury* (lying under oath) Americans were dismayed. It really did seem that the State Department might be full of spies and traitors. Alger Hiss spent four years in jail. (Hiss was guilty of lying, but was he a spy? People still argue about that.) Of course, everyone knew that Russia had spies in the United States and that we had spies in Russia. Nations spied on each other then. They still do.

No question about it, these were confusing and frightening times. The United States had believed it was alone in having atomic power. Then, soon after the war's end, Russia tested an atom bomb. The thought of Joe Stalin, a coldblooded tyrant, with an atom bomb was terrifying. (It became still more terrifying when both nations developed hydrogen bombs.)

And there was China. For centuries, China was under the rule of corrupt warlords. The Chinese people wanted a better government, so, even before World War II, they looked to other leaders. That led to a civil war, with two groups fighting for control of the huge country. One group, led by Chiang Kai-shek, was known as the Nationalists. The others were the communists, led by Mao Zedong.

During the Second World War, most Chinese fought together

Congressman Richard Nixon shows off some microfilm, part of the evidence used to convict Alger Hiss (right) of perjury in 1950.

39

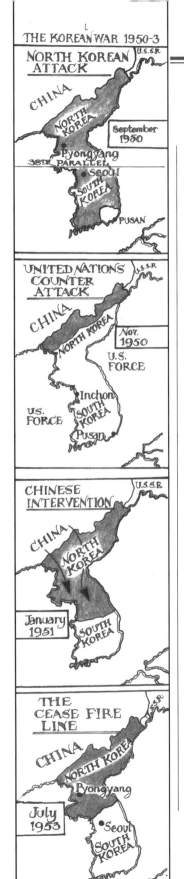

against their common enemy—Japan. But as soon as the war was over, they were fighting among themselves again. Finally, the communists won. But things didn't work out as most people expected. China's communists brought land reform and some stability to the country, but they also brought unfree totalitarian government. In America, people with loud voices said that Harry Truman was to blame for China's new communist government. (This may not seem to make sense now. Well, it didn't then, either, but some people listened.) Newspapers carried angry letters about China and how "we" lost it.

We had had some unusually well-trained China experts attached to our State Department. When those experts predicted that China would fall to the communists, the anti-communists in America didn't want to hear that news. America's China experts were accused of being communists themselves. They were fired, and replaced by others who were strangers to the region. That left us with almost no experts in the difficult years that were to come. Because of poor advice, we would make some bad mistakes in East Asia, especially in a place called Vietnam.

In 1950, however, we acted boldly. Most people think we did the right thing when the ruler of North Korea sent a powerful army into South Korea. Look at the maps and you'll see Korea, the big peninsula that juts out into the sea between China and Japan. Korea was an ancient, independent country. It was divided in two at the end of World War II. It was

By the end of 1949, Chiang Kai-shek, the former U.S. ally, had been driven off mainland China—to the island of Formosa (now Taiwan)—by Chinese communist forces, led by Mao Zedong (above, communist troops enter the city of Nanking).

divided at latitude 38° north, also known as the *38th parallel*. Korea was supposed to be brought back together with free elections. The communists—who controlled the north—never allowed those elections.

The Russian-educated North Korean leader, Kim Il Sung, intended to make all of Korea communist. His army had the latest Russian tanks and equipment.

The leader of South Korea, Syngman Rhee, had been educated in America. His army was poorly trained and badly equipped.

When North Korea's army entered South Korea, it was a test for the world community, and for Truman and his policy of *containing* communism (keeping it from expanding into new regions). The world had stood by and let Hitler conquer weaker nations. This time there would be no giving in. The United Nations acted quickly. The United States led the response. The Cold War had turned hot. The Korean War had begun.

The fighting in Korea was tough. At first our soldiers took a beating. Then things changed; we began to win. And then the commander of our troops in Korea got carried away. He was General Douglas MacArthur, and he had a big ego and not much respect for President Truman. MacArthur was a great general when it came to strategy, but he disobeyed orders to stay in South Korea, and, while his troops were winning, he went beyond the 38th parallel and into North Korea. That brought communist China into the war (to help its friend Kim Il Sung), and that changed everything. The Chinese sent highly trained, well-equipped troops. The war turned again. It was a terrible situation. We were no longer just at war with Korea. Truman and his advisers feared that this might be the start of World War III. The fighting kept zigzagging back and forth. Everyone had expected a quick war, but now it seemed as if the Korean War would never end.

The American people were jittery and worried. We were fearful that communism would dominate the world. Because of those fears, we did some foolish things at home. We lost faith in ourselves. Some people thought that communists were about to take over the United States. There was no good reason to believe that, but fear often isn't reasonable. So we Americans took part in (or kept quiet during) a communist hunt in the United States. We persecuted some of our own citizens. It was a time of panic—as bad as, or maybe worse than, the witch hunt in Salem, Massachusetts, in 1692. And that was mighty hard on the Puritans.

Early in the war, U.N. forces were pushed out of Pyongyang, North Korea's capital, back over the 38th parallel into South Korea.

In 1949, the United States and ten European nations formed the North Atlantic Treaty Organization (NATO). It was to be an instrument of defense against the threat that they felt the Soviet Union and its eastern European satellite countries posed to the West.

General MacArthur visits the front. Truman eventually fired him for disobeying orders.

8 Tail Gunner Joe

The Statue of Liberty? No —it's Joe McCarthy, doing his best to put out America's torch of freedom.

Disregard for facts? Accusations without proof? Half-truths that confuse? Do you think anyone does those things today? (Can you find some examples?)

He was a liar. Not your ordinary small-time fibber. No, Senator Joseph McCarthy was an enormous, outrageous, beyond-belief liar. Trouble was, some people believed him. After all, no one thought a United States senator would lie.

Of course, most people didn't realize he had lied to become a senator. He said he was a big war hero, a tail gunner who had shot down lots of enemy planes. Actually, he had spent most of the war at a desk job. And he made up stories about his opponent, Wisconsin senator Robert M. La Follette, Jr. Those stories helped him beat La Follette. Washington's newspaper reporters had voted Robert La Follette the best of all the senators; they soon voted Senator Joe McCarthy the worst.

Joseph McCarthy was a man who liked attention. He wanted people to notice him; he needed an issue. What was an issue that would capture headlines? Communism. Some people were afraid that America might become a communist nation. Joe McCarthy would tell them the danger was real. He would scare them. At a speech in Wheeling, West Virginia, he waved a piece of paper and said it contained the names of 205 communists who worked in the U.S. State Department (the State Department runs our foreign policy! Alger Hiss had been in the State Department!).

McCarthy was lying about the list, but many Americans believed him. That was the beginning of McCarthy's witch hunt. Before he was finished he had accused hundreds of people of communist activity.

He never proved a single case against even one person. But it didn't seem to matter. He was a master of publicity. McCarthy was an exciting speaker. His accusations captured people's attention. Whatever he said got printed in bold headlines. And he knew how to use that new medium

"I have here in my hand a list of 205 names," said McCarthy. But he never actually showed the names to anyone.

Left: the chairman of the House Un-American Activities Committee at a hearing on supposed communist activity in the film industry. Movie star Gary Cooper (right) took the stand and declared that he was against communism. "From what I hear," he said, "it isn't on the level."

—television. He was almost like a circus performer putting on a show.

Joe McCarthy wasn't entertaining to the people he accused. They lost their jobs. Some lost their homes. Often their friends deserted them. Their lives were ruined.

The nation was infected with a bad case of anti-communist hysteria. It was sick. What about free speech and the guarantees of the First Amendment? Those rights were in trouble. During the McCarthy era, most people were afraid to speak out. It was a time of great fear.

McCarthy wasn't the only one who ignored the Bill of Rights and its protections. The House of Representatives had a committee called the House Un-American Activities Committee. It ruined lives, too. HUAC decided to investigate the movie industry; many people had the idea that artists and actors were likely to be communist sympathizers.

Anyone questioned by the committee might be put on a list, called a *blacklist*. People on the blacklist—actors, producers, writers, cameramen—were unable to get jobs in the film industry. Some filmmakers had been members of the Communist Party; some had attended communist meetings, often out of curiosity, or out of Depression-time fear that capitalism was doomed. People called before the congressional committee were asked to name others who had attended those meetings. Anyone named would probably be put on the blacklist. Some writers and actors refused to answer HUAC's questions; many were sent to jail.

To repeat: it was a time of great fear. Ordinary people in America were afraid to buy books, subscribe to magazines, or join organizations that might have the slightest left-wing leanings. Lots of people believed McCarthy's baseless lies. Fear of communism muddied their thinking. But in a free country it is not a crime to hold any kind of belief—and that includes communist beliefs.

The anti-communist extremists wanted to prevent people from reading about communism. They wanted to make it a crime to be a communist. McCarthy made a list of 418 American authors who he said had disloyal ideas. His list included great writers like Ernest Hemingway, John

Communism is a philosophy of the extreme left; HUAC and McCarthy represented the extreme right. Sound thinking is usually in the middle.

Communists (and socialists, or anyone politically left-wing) were often called names (usually rudely) like *reds*, *pinks*, or *pinkos*.

Hysteria is an exaggerated emotional reaction to fear, horror, or disgust.

When McCarthy's campaign began, the U.S. Communist Party was a legal organization—as it is today.

Communists do not believe in God. That was one reason they were disliked, even feared by some Americans. Writing about religious freedom, Thomas Jefferson said, "The legitimate powers of government extend to such acts only as they are injurious to others. But it does me no injury for my neighbor to say there are twenty gods or no god. It neither picks my pocket nor breaks my leg."

Anti-communism wasn't limited to the movies. In 1950, Chrysler autoworkers in Los Angeles beat up a fellow worker for refusing to say if he belonged to the Communist Party.

Dos Passos, and Henry David Thoreau. The State Department removed their books from overseas libraries (which were the only libraries it controlled).

Thirty-nine states passed anti-communist laws. Texas made membership in the Communist Party a crime punishable by 20 years in prison. A Connecticut law made it illegal to criticize the United States government or flag. Loyalty oaths were demanded of government workers, including many teachers! The oaths varied—generally the jobholders had to say that they supported the government or were not communists. (Ask your teachers what they think about this.)

If HUAC's members had read the Bill of Rights, they had forgotten what it says. They didn't seem to understand that our Founders had challenged us to do something very difficult: to provide free speech to all, including those whose ideas we detest. Thomas Jefferson and James Madison and the other Founders thought America's citizens should be free to examine any idea, including ideas most people find loathsome. Now that isn't easy at all. You can only do that if you really believe in Jefferson's words that "truth is great and will prevail if left to herself."

Many of our government officials turned cowardly during this time of great fear. Some hated communism so much that they didn't seem to care if people's lives were destroyed. Many really did think the government was full of communists. Others were scared. They had reason to be scared. Two senators who opposed McCarthy were defeated. Eight senators whom McCarthy supported were elected.

But some Americans had courage. Senator Margaret Chase Smith of Maine spoke out in Congress.

I think it is high time that we remembered that the Constitution speaks not only of the freedom of speech but also of trial by jury instead of trial by accusation....I am not proud of the way we smear outsiders from the floor of the Senate and...place ourselves beyond criticism.

Edward R. Murrow, a television newsman (who had broadcast on the radio to America every night from London when it was being bombed during the Second World War), was another with courage and integrity too. He decided to give the American people a clear picture of Joe McCarthy. Murrow made a TV film that showed the senator yelling at

Detest means to loathe or hate.

The real scandal in all this [the McCarthy era] was the behavior of the members of the Washington press corps, who, more often than not, knew better. They were delighted to be a part of his traveling road show, chronicling each charge and then moving on to the next town, instead of bothering to stay behind to follow up. They had little interest in reporting how careless he was or how little it all meant to him. It was news and he was news; that was all that mattered.

—DAVID HALBERSTAM, *THE FIFTIES*

COMMUNIST PARTY ORGANIZATION U.S.A–FEB. 9, 1950

At the Army–McCarthy hearings, McCarthy points to a map purporting to show the location of communists. Joseph Welch (left), the army's lawyer, turned to him: "Until this moment, Senator, I think I have never gauged your cruelty or recklessness," he said. "Have you no sense of decency?"

witnesses, belching, picking his nose, and making contradictory statements. Murrow knew he might get blacklisted himself, but that didn't stop him. He said, "This is no time for men who oppose Senator McCarthy's methods to keep silent." Then he added, "We must always remember that accusation is not proof."

McCarthy kept attacking. He accused the U.S. Army and many of its officers and soldiers of being communist sympathizers. But he produced only an army dentist who may, at one time, have been a

Margaret Chase Smith

communist sympathizer. Television (which was a new addition in many American homes) let people see the Army–McCarthy hearings and the senator's shouting, sneering manners.

Most people were disturbed by what they saw. But that didn't stop Joe McCarthy. He went on smearing innocent people.

Then a quiet, elderly man, who was well respected in his home state of Vermont but little known elsewhere, spoke out in the Senate.

Joe McCarthy, G-Man

Two famous journalists, the brothers Joseph and Stewart Alsop, visited McCarthy at his headquarters in the basement of the Senate Office Building and described what they saw:

A visit to the McCarthy lair on Capitol Hill is rather like being transported to the set of one of Hollywood's minor thrillers. The anteroom is generally full of furtive-looking characters who look as though they might be suborned State Department men. McCarthy himself, despite a creeping baldness and a continual tremor which makes his head shake in a disconcerting fashion, is reasonably well cast as the Hollywood version of a strong-jawed private eye. A visitor is likely to find him with his heavy shoulders hunched forward, a telephone in his huge hands, shouting cryptic instructions to some mysterious ally.

"Yeah, yeah, I can listen, but I can't talk. Get me? You really got the goods on the guy?" The senator glances up to note the effect of this drama on his visitor. "Yeah? Well, I tell you. Just mention this sort of casual to Number One, and get his reaction. Okay? Okay. I'll contact you later."

The drama is heightened by a significant bit of stage business. For as Senator McCarthy talks he sometimes strikes the mouthpiece of his telephone with a pencil. As Washington folklore has it, this is supposed to jar the needles of any concealed listening device.

I Will Not Cut My Conscience to Fit This Year's Fashion

Playwright Lillian Hellman was called before the House Un-American Activities Committee. It was 1952, and she sent this letter to the committee chairman:

Lillian Hellman

Dear Mr. Wood:

...I am most willing to answer all questions about myself. I have nothing to hide from your Committee and there is nothing in my life of which I am ashamed....But I am advised by counsel [lawyers] that if I answer the Committee's questions about myself, I must also answer questions about other people....This is very difficult for a layman to understand. But there is one principle that I do understand: I am not willing, now or in the future, to bring bad trouble to people who, in my past association with them, were completely innocent of any talk or any action that was disloyal....I do not like...disloyalty in any form, and if I had ever seen any, I would have considered it my duty to have reported it to the proper authorities. But to hurt innocent people whom I knew many years ago in order to save myself is, to me, inhuman and indecent and dishonorable. I cannot and will not cut my conscience to fit this year's fashions....I was raised in an old-fashioned American tradition and there were certain homely things that were taught to me: to try to tell the truth, not to bear false witness, not to harm my neighbor, to be loyal to my country, and so on....It is my belief that you will agree with these simple rules of human decency and will not expect me to violate the good American tradition from which they spring....I am prepared...to tell you everything you wish to know about my views or actions if your Committee will agree to refrain from asking me to name other people....

Sincerely yours,
Lillian Hellman

Lillian Hellman wrote a book about the McCarthy era and called it Scoundrel Time. *Who do you think Hellman thought were the scoundrels? Do you agree with her? Do you think there are any scoundrels in politics today? It isn't enough to come up with names—you have to have a good reason for calling someone a scoundrel.*

His name was Ralph Flanders, and he said of Joe McCarthy:

He dons his war paint. He goes into his war dance. He emits war whoops. He goes forth to battle and proudly returns with the scalp of a pink dentist.

Flanders asked the Senate to vote to censure (condemn) Joseph McCarthy. One senator immediately said that Flanders must be on the same side as the communists.

But by this time most Americans had had enough of Joe McCarthy. Murrow's film had shocked them. Senators began hearing from the voters; most of them were tired of the witch hunts. The Senate voted on Ralph Flanders's measure: Joseph McCarthy was censured for outrageous behavior. The man who had ruined lives and terrorized much of the nation was disgraced. However, he remained in the Senate. (McCarthy died a few years later from a liver ailment caused by too much drinking.)

Neither the executive branch, nor the legislative branch, nor the judicial branch of our government acted boldly during the time of the communist fear. Later, most Americans were ashamed of the McCarthy witch hunts. Joseph McCarthy and HUAC made us aware of the preciousness and fragility of our right to free speech.

Freedom of speech is guaranteed by the Bill of Rights. It is your constitutional right as an American citizen. But that freedom to speak out is easy to attack in times of crisis. It takes citizens who appreciate its importance to make sure we keep it as a basic right. And don't forget, the right of free speech also belongs to those whose ideas you hate.

Fragile means delicate or easy to break.

9 Liking Ike

Eisenhower was nominated as Republican presidential candidate at the first national convention covered on TV.

Some people called them "the nifty fifties" and said it was a glorious time. After all, there was a singer named Elvis Presley, the beginning of rock and roll, two new states (Hawaii and Alaska), hula hoops, a movie star named Marilyn Monroe, the Salk vaccine (which prevented polio), and television. TV wasn't new. It had been invented in the '30s. Back then, a Harvard expert said it would never make an impact like radio, because "it must take place in a semi-darkened room, and it demands constant attention."

By 1946, a few people were willing to pay attention. That year 7,000 small, black-and-white TV sets were sold in the U.S. and regular programming was underway. By the mid-'50s, more than 5 million TV sets were sold each year. In 1956, videotape was developed. That meant shows could be taped, edited, and rerun. Before videotape, all TV was "live." If an actor made a mistake, everyone saw and heard it.

Two southern Californians made a plastic version of a bamboo toy they'd heard that some Australian children were playing with. By 1958, they'd sold 25 million hula hoops.

By 1949, Americans were buying 100,000 TV sets a week; quiz shows and soap operas were popular. The most watched program by kids under 13 was the cowboy show *Hopalong Cassidy*. Television brought a new audience to professional sports such as baseball, boxing, and football.

Marilyn Monroe sometimes played a dizzy, dumb blonde—but she was neither dumb nor dizzy. Rent a video of one of her movies, *Some Like It Hot.* You will laugh until you're tired.

Joseph Stalin died in March 1953, two months after Eisenhower took office as president. Many Russians—who didn't know about all the terrible things Stalin had done and all the innocent people he had killed—wept as though their own father had died.

In 1950, 90 percent of America's homes did *not* have TV. By 1960, 90 percent of America's homes did have it. That magic box brought the whole world into living rooms from Honolulu to St. Paul to Miami. TV wasn't just a luxury for the rich; it was very democratic. Everyone had TV: rich and poor, city folks and country cousins. Everyone saw the same events and laughed at the same comedians. It gave us a common culture. But some people asked: what kind of culture is it?

About a third of the children's programs featured crime and violence (although 1950s violence seems tame today). TV became a baby-sitter—it kept the kids quiet—and fewer parents took time to read to their children. Families ate their dinners in front of the TV set, and talked to each other less. Television changed the way politicians campaigned: there were no more Harry Truman whistle-stop train trips. Before long, the candidates—packaged by makeup artists and TV coaches—were coming right into the living room.

But all that was just beginning in 1952, when we elected a new president, our 34th. His name was Dwight D. Eisenhower, and he was immensely popular. People called him "Ike." Eisenhower, an army general, had been Supreme Allied Commander (which means he was head man) in Europe during World War II. He had light blue eyes, a balding head, a grandfatherly manner, and the friendliest grin you can imagine. His campaign buttons said I LIKE IKE—and most people did.

He was so popular that Democrat Harry Truman asked him to run for president. But Eisenhower was a Republican, so he ran against Truman's party. After 20 years of Democratic leadership, Americans were ready for a change. The Democrats had been the party of active government. Franklin Delano Roosevelt was a strong, dynamic president who gathered idea people around him; he brought college professors into government. Harry Truman was naturally combative; he liked to confront problems and make decisions.

Eisenhower's style was very different. He was a conservative, and

In 1954, Dr. Jonas Salk (right) prepares to inject a little girl with his polio vaccine. Polio and diphtheria vaccinations hugely reduced the effects of those crippling diseases.

most of his advisers were businessmen. He believed the president should be a strong moral leader. But he didn't think he should take sides.

Ike believed in persuasion and patience. He thought the president should act quietly. Eisenhower put a sign on his desk. It was in Latin but, translated, it said: *gentle in manner, strong in deed.* Which was exactly how he tried to be. Eisenhower promised stability, and as little government action as possible. He played golf and always seemed relaxed. His critics called him the "stand-still" president. They thought he was lazy. They were wrong.

Eisenhower worked hard and held the reins of the presidency tightly. But he tried to give the appearance of being above the political battle. He believed in behind-the-scenes leadership. Eisenhower didn't think the president should be controversial. So he didn't speak out against Joseph McCarthy, even when McCarthy outrageously criticized his friend General George Marshall. But he did fly to Korea, as he had promised when he campaigned for the presidency.

Eisenhower wanted to end the Korean War. Sometimes it takes strength to quit a fight. Eisenhower had that strength. He didn't want any more deaths; he didn't want to risk war with China.

President Eisenhower knew the waste of war. He had seen it

World War II British general Bernard Montgomery said of Ike, "He merely has to smile at you and you trust him at once."

The Latin motto that Eisenhower kept on his desk was *suaviter in modo, fortiter in re.*

About 54,000 Americans died in the Korean war. We never officially declared war on Korea and there was no peace treaty.

Blustering Khrushchev

Khrushchev was a boisterous miner's son who once took off his shoe at a United Nations session and banged it on the table to make a point.

Nikita Khrushchev became the first secretary of the Soviet Communist Party in 1953 (we say KROOSH-chev, but the Russian pronunciation is more like hroosh-CHOFF. The *kh* sound is like the *ch* in *chutzpah*—pronounced HOODZ-pah—which Khrushchev had a lot of). He didn't really say "We will bury you" (see page 51). The translator got it wrong. Khrushchev actually said something like "We will leave you in the dust." But with all the Cold War hysteria, those wrong words got repeated. Fear was in the air; the arms race took off.

As Eisenhower viewed the situation...the possible menace of the Soviet Union took two forms. One was the external threat of Soviet military might. The other was internal—in the sense that the existence of this power...might drive the United States into weakening and eventually destroying its own economy through the indefinite expense of preparedness. What he proposed, therefore, was that the United States should strive for a middle road.

—ROBERT J. DONOVAN, *EISENHOWER, THE INSIDE STORY*

with his own eyes. "Every gun that is made, every warship launched," said Eisenhower, "is a theft from those who hunger and are not fed, from those who are cold and are not clothed."

The Korean War came to an end. No one won. Korea was left divided as it had been when the war began. But the United States and the United Nations had proved what they had set out to prove: they would stand up to communist aggression.

Eisenhower saw a danger ahead for the nation. It was something he was an expert on: military power. In his farewell address he warned of a phenomenon "new in the American experience...an im-

Many liberals felt that Eisenhower didn't seem to care about social problems at home—especially the problems of civil rights, which were becoming hard to ignore. More about that later in this book.

The King

Elvis was called the King of Rock, and that is exactly what he was. He was a white boy who sang black music with tremendous natural talent and energy. He made it acceptable to white listeners. Elvis said, "The colored folk been singin' it and playin' it just the way I'm doin' now, man, for more years than I know." In terms of popular success, no American musician could touch him. Elvis was polite and lonely; he didn't smoke, he didn't drink, and he didn't eat meat, but many adults found his singing dangerous. It was loud, and that *was* dangerous to eardrums. But it was the way he wiggled around that upset some people. They'd never seen anything like it. America's youth fell in love with him. Elvis didn't have much to say. He had his talent, and he liked making money and he made a whole lot of it. But he didn't know how to handle fame. All the money and attention changed the clean-cut young man into a dissipated drug taker. Elvis was killed by years of overeating and overindulgence in drugs.

From 1956 to 1958, Elvis made 14 consecutive million-selling records, starting with "Heartbreak Hotel"—and broke hearts wherever he went.

By the middle of 1951, the end was in sight in Korea, and troops such as these marines were pushing North Korea's forces back. But country and people had suffered terribly—right, a South Korean child during the invasion of Inchon.

mense military establishment and a large arms industry." He called those two groups (the military and the weapons makers) the *military-industrial complex*. He was prescient (PRESH-unt), which means he was seeing the future.

As soon as Eisenhower was out of office, an arms race began in earnest. It was a contest with Russia to see who could build the most weapons. And Republicans and Democrats jumped on the arms wagon. Both our nation and the Soviet Union spent—and went on spending—vast amounts of capital to build guns, bombs, and missiles. It was a theft from those who hungered and were not fed.

Stalin was dead, and no one in the United States knew quite what to make of the new Soviet leaders. But when Premier Nikita Khrushchev said, "We will bury you," it didn't sound friendly. So most Americans supported the arms race—which meant more and more weapons, more and more military, for more and more money, which put us further and further in debt. We didn't listen to President Eisenhower. He, a former general, had reduced military spending.

Good Times?

The Eisenhower years were prosperous. Jobs were plentiful. People had money to spend. During World War II, there had been shortages of consumer goods. Now you could buy bikes, vacuum cleaners, television sets, dishwashers, ballpoint pens, nylon stockings—almost anything you wanted. For three cents you could mail a letter; for five cents you could buy a Coke or a candy bar. There were no big fast-food chains, although they would come soon enough. (Five cents for a Coke wasn't quite as inexpensive as it sounds. Salaries were a whole lot lower then.)

These were good times.

But not for everybody. Some citizens were kept out of the good times. In the South, blacks couldn't eat in the same restaurants as whites, shop in the same stores, use the same bathrooms, drink from the same water fountains, or go to the same schools. It was humiliating—and unfair. But it was the law. It was all because of a Supreme Court decision, back in 1896, called *Plessy* v. *Ferguson*. That decision said that as long as facilities were equal, they could be separate. It made segregation legal. But the segregated schools, restaurants, and shops weren't equal; anyone could see that. Even if they were, who wants to be separated? Americans were supposed to be free. How can you be free when you can't go where you want to go?

10 Houses, Kids, Cars, and Fast Food

Thirty million "war babies" were born between 1942 and 1950—and that was just the beginning of the baby boom. These ones were competing in a "Diaper Derby."

There were suburbs before World War II. But they were mostly pretty communities for the very wealthy. The exodus from the city to modern suburbia was a mass movement.

Couples had put off having children during the war years—and now they were making up for that. We were having a "baby boom." The war veterans had gone to college under the G.I. Bill of Rights, and the government paid for their tuition. By the '50s, most of those veterans were out of school, married, and having children. Their college degrees helped them find good jobs, usually better jobs than their fathers had ever had (most women—at

A family and their brand-new Levittown house. Every lot measured 60 feet by 100; every house had 720 square feet. Right, Levittown from the air.

52

least middle-class women—didn't work).

Those new families needed places to live, and, in America, every family dreamed of a home of its own. But there was a big housing shortage. What to do? Use some American ingenuity.

William Levitt had it. Before the war, an average builder might build two or three or, at most, five houses a year. Bill Levitt was soon finishing 36 houses a day, which added up to 180 in every five-day week! How did he do it? By analyzing the building process, dividing it into 27 steps, and putting teams of people to work on each step. It was Henry Ford's mass-production idea applied to housing. A team did the same task, over and over, moving from house to house. There were framers and roofers, tile men and floor men, painters who did all the white painting and others who painted all the green. If anyone slowed down, it fouled up the whole production process. Bill Levitt made sure that didn't happen. He began producing his own nails and making his own cement. He even bought timberland in Oregon and cut his own lumber. By doing all that, he kept his house prices very low.

He had thought all this out while he was in the navy, where he was assigned to the Seabees. (They were the navy's builders.) Levitt was commissioned to build airfields, practically overnight. Lives depended on his speed. He analyzed, planned, brainstormed with other Seabees, and built the airfields. Later, he said the navy gave him a chance to experiment and learn how to get things done.

Levitt knew that a lot of veterans like himself would be looking for homes after the war. So he bought a huge tract of land on Long Island (near New York City). It was mainly potato fields. Those fields soon became a community called Levittown. Most of Levitt's houses had four and a half rooms and were exactly alike. They were sturdy, available, and a great value—like Ford's Model T. When the first advertisement for the first Levittown ran in the *New York Times*, people began lining up. In one day alone, Levitt sold more than 1,400 houses. Bill Levitt's ideas were soon copied by other builders. The communities they built were part of something that was about to boom: *suburbia*. Suburbs—on the outskirts of the cities—were springing up around the country. Some had low-cost houses, but others were for the affluent. As people moved out of cities, new people—often poor people—moved in. Cities began losing some of their most productive taxpayers just when they needed rebuilding.

In the new suburbs, where there was no mass transportation, something became essential: a car (or two). Well, General Motors (and the

Hula hoop and Frisbee fads were followed by a rage for 3-D comics that you read with special glasses.

What were people watching on their new television sets in the 1950s? The A. C. Nielsen company, which measures TV watching, says that the number-one show for most of the decade was *I Love Lucy* (more about Lucy in Chapter 29). Other big shows were *Gunsmoke*, the *Ed Sullivan Show*, *Dragnet*, the *Jackie Gleason Show*, *Wagon Train*, *You Bet Your Life*, and *General Electric Theater*. The host of *GE Theater* was an actor named Ronald Reagan.

Hanging out, 1950s style, at the record store (this one was in Webster Groves, Missouri). More listening went on than buying. The uniform (for girls) was pleated skirt, sweater, bobby socks, and saddle shoes.

Two monster symbols for the '50s: a 1957 Chevy Bel Air—whose engine could propel the car at least twice as fast as any speed limit allowed it to go—and an outdoor drive-in movie theater, where you and your sweetheart sat in your car to watch the show.

other auto companies) saw to that. Detroit (the home of the auto industry) began building big, fancy cars. They were status symbols with shiny chrome trim, shark-like tail fins, and seductive shapes. The advertising industry geared up to convince consumers that last year's car, like the hem length on last year's dress, was out of date. Styling changes, not engineering excellence, determined the big sellers. Raymond Loewy, a famous industrial designer, said the showy '50s cars looked like "jukeboxes on wheels."

But American consumers didn't care; they liked jukeboxes. To most people, bigger seemed better. And bigger and fancier seemed best of all. These young '50s families didn't seem to have a problem paying for the new houses and the big cars. We were becoming an affluent nation. In the old days, most workers had toiled for low wages to enrich factory owners. Now, new union contracts were giving auto workers and others a share of their own productivity. They became some of the best customers for the cars and appliances that were rolling out of the factories. And they set the standards for worker pay in other fields too.

During this same time, oil was replacing coal as our major source of energy. In 1949, Americans used 5.8 million barrels of oil a day; by 1979 we were using 16.4 million barrels. Oil is much more efficient than coal. Getting it doesn't demand exhausting, killing work. (How would you like to dig coal in a coal mine?) And oil was very cheap. (That changed a few decades later.) Oil began driving the economy. It allowed Americans to get into their big new cars and not worry about how much gas they guzzled.

With a family and a car, chances are you'd want to take a trip. Kemmons Wilson, who was a house builder in Memphis, Tennessee, did just that. It was 1951, and he decided to take his family to Washington, D.C., to see the sights.

You know how kids sometimes behave in the car, don't you? Well, the Wilsons' children were no different from anyone else's. Kemmons and his wife couldn't wait each night to get to a motel and relax.

But the motels they found were mostly either disappointing or awful—and they charged extra for each child in the room, even though the Wilsons' kids had brought their own sleeping bags!

Kemmons Wilson thought about those motels while he was in Washington, and then he couldn't wait to get back to Memphis. He had an idea buzzing in his brain. He was going to build motels for families. The motels would all be similar, so people would know what they were getting. They would be clean and attractive. "And," said Wilson, "if I never do anything else worth remembering in my life, children are going to stay free at my motels."

Popular singer Bing Crosby had made a movie called *Holiday Inn*. Wilson thought it a great name for a motel. In 1956, when Congress passed a huge $76 billion federal highway program, Kemmons Wilson and his Holiday Inns were ready for all the traffic those new highways brought.

Cars, suburbs, and TV watching were changing American habits. But some things hadn't changed much. Most American families never went out to dinner. It was too expensive to eat out with the kids—unless it was a very special occasion. Restaurants were apt to be costly, or sleazy. Mothers—especially those suburban moms—were expected to stay home and cook meals.

A lot of working people didn't go out either. They packed lunch in a sandwich bag and brought it to work. Two California brothers were going to change that. They were going to build a restaurant that was fast, clean, and very inexpensive. It was a place where you could feed the whole family and your wallet wouldn't be wiped out.

Dick and Mac (Maurice) McDonald had come to California in the Depression '30s. They wanted to be movie producers. But it didn't happen. So they opened a small movie theater. That folded. Then, in 1940, they built a drive-in restaurant. People stayed in their cars at the drive-in and waiters, called carhops, came out and served them. The restaurant

Kemmons Wilson's first Holiday Inn, in his hometown of Memphis, Tennessee. It opened in 1952; by 1978, Memphis *alone* had 15 Holiday Inns.

A carhop serves moviegoers at one of the 2,000 drive-ins built between 1947 and 1950.

55

Dick and Mac's original McDonald's, in San Bernardino, California, in 1955. "We have sold over 1 million," the sign said. What do the McDonald's signs say today?

Suburbia has become the quintessential physical achievement of the U.S.: it is perhaps more representative of its culture than big cars, tall buildings, or professional football.
—KENNETH T. JACKSON, *CRABGRASS FRONTIER*

appealed to families, especially during the war years, when lots of women went to work and didn't have time to cook.

But the McDonald brothers had a passion for efficiency, and cars were lining up. People were waiting to get service. How could they speed things up?

Well, the carhops would have to go. Then the McDonalds looked at their menu. There was too much choice—that slowed things down. And it took time for the customers to put ketchup and relish on their hamburgers. Besides, the condiment stand was always messy. The McDonalds hated mess. They decided to put pickles, mustard, ketchup, and onions right on the hamburgers. That really saved time. Instead of regular dishes and silver they changed to paper plates and plastic forks. That was faster. But what was most important was the way the kitchen was organized. They set up a production line. They even invented and adapted their own kitchen equipment. It was the Henry Ford idea again, this time applied to hamburgers. Grill men cooked the burgers, milkshake men made shakes, dressers wrapped the burgers, and countermen took orders. By the middle '50s, people were lined up to eat burgers at the one McDonald's restaurant; it was in San Bernardino, California. The brothers were rich, and happy. They didn't need any more money than they had.

But some people wanted to open McDonald's restaurants in other places. They wanted to buy *franchises* (that means they were willing to pay for the McDonald name and expertise). The brothers didn't want the bother.

One day, Ray Kroc came by. He sold milkshake makers, and he wanted to see why the McDonalds were buying so many of them. He was astounded by what he saw: long lines of people waiting to buy hamburgers. The place was spotlessly clean, the hamburgers were good, and they cost 15 cents. Dick and Mac McDonald were looking for someone to handle their franchising. Kroc was eager. A few years later, he bought the McDonald's name and idea outright. He was on his own and ready to make business history. Kroc was 52 and he had health problems, but that didn't stop him. He was a workaholic and a perfectionist. The McDonald brothers were amazing; Ray Kroc was more so.

He began opening McDonald's hamburger stores one after another. Soon they were ubiquitous (yoo-BIK-wit-us), which means they were everywhere. He wanted the hamburgers he sold in Des Plaines, Illinois, to be exactly like the hamburgers he sold in Willmar, Minnesota, or

Kalamazoo, Michigan. He wanted every restaurant to be immaculate and to maintain high standards. He made rules, lots of rules. McDonald's workers couldn't have beards or mustaches, and they couldn't chew gum. He made sure their fingernails were clean. He didn't like to hire women; he thought they would flirt with the customers. He worked very, very hard. Even when he became enormously successful he never put on airs. When the company grew to be huge, with a McDonald's in almost every town and village, and his worth in the hundreds of millions, he still insisted that his executives answer their own phones. No high hats for him, he said. It wasn't money that interested him. "I worked for pride and accomplishment," he said. "Money can be a nuisance. It's a hell of a lot more fun chasin' it than gettin' it. The fun is in the race."

His success was related to the new way of life in America. McDonald's were suburban restaurants, they took advantage of locations on the new highways, and, as more and more women began to get jobs, dinner at McDonald's—or at one of the other fast-food chains that followed—became a regular thing.

Suburbs are small, controlled communities where for the most part everyone has the same living standards, the same weeds, the same number of garbage cans, the same house plans, and the same level in the septic tank.
—ERMA BOMBECK, HUMORIST

Families lined up to buy identical suburban houses; then they tried hard to make each one as different from the others as possible.

Ah, Suburbia, Happy Suburbia

Nineteen-fifties suburbia—or its image, anyway—was part Hollywood movie, part TV comedy series, and part slick magazine ad. There was Mom (who was pretty), Dad (who was handsome), two kids (who were cute), and Rover (who barked appealingly). A station wagon sat in the driveway. (The garage held a lawnmower, bikes, Dad's workbench, roller skates, the family camping gear, and had a basketball hoop above the door.)

Everyone was white. (Black people couldn't buy houses in Levittown, or most of the other early suburbs. That changed later.) Everyone was young. (Grandma and Grandpa were back in their small hometown, which was getting smaller as young couples moved to suburbia.) All the families earned about the same income. (If they began earning more, they usually sold the house and moved to a fancier suburb.)

Always before, communities had people of all ages to balance each other. In suburbia, children missed having grandparents nearby. And separating the races (blacks in the cities, whites in the suburbs) was not going to lead to harmony between them. Those picture-pretty suburban families were expected to conform; any who didn't felt out of place. Mothers weren't supposed to work—it spoiled the picture. But some women were bored; there wasn't a whole lot to do in the suburbs—except drive the kids here and there. And Dad got to hate the commuting and the highway traffic. Suburbia wasn't perfect, even though the magazine ads made you think it must be.

11 French Indochina

Ho Chi Minh's name means *enlightened leader of the Vietminh*. He changed his name (from Nguyen Sinh Cung) to reflect his total and lifelong devotion to the fate of his country.

There was a beautiful country to the south of China. A country that had been ruled by France for about 100 years. The French called the land *Indochina*. The people who lived there called it *Vietnam* and *Laos* and *Cambodia*. Does that sound unusual, for a Western nation to rule an Asian nation? It wasn't unusual at all in the days before World War II. Britain ruled India. The Netherlands ruled Indonesia. The United States ruled the Philippine Islands. That kind of foreign rule is called *colonialism*, or *imperialism*. Most people knew that

colonialism wasn't fair. (We had once been an English colony and we hadn't liked it a bit.) Franklin Delano Roosevelt said that nations should determine their own form of government. As soon as World War II ended, we granted independence to the Philippines, but some Asian countries had to fight to become free. India's leaders fought with hunger strikes and nonviolent protests. In Vietnam they fought with weapons.

Vietnam had a leader named Ho Chi Minh. As a young man, he had visited in England, the United States, and France, and he knew Western ways. He'd been a poet and a photographer, and he spoke many languages. Back in 1919, he had asked President Woodrow Wilson to help the people of southeast Asia gain their freedom. But Wilson had not responded.

When Ho learned of Philippine independence he was encouraged. He wrote eight letters to President Truman asking for help in making Vietnam free. Those letters were never answered. Perhaps Truman never saw them. We don't know about that, but we do know that in 1945 Ho Chi Minh founded the Democratic Republic of Vietnam. Some American military men were present at the independence ceremonies. "The Star-Spangled Banner" was played by a Vietnamese band, and Ho Chi Minh spoke words from the Declaration of Independence. When American planes unexpectedly flew overhead, everyone cheered.

But the French didn't want Vietnam to be independent. France had lost the country when Japanese soldiers invaded during World War II. Now that the war was over, the French wanted their old empire back. Ho Chi Minh stood in the way. Ho was a hero to his countrymen. He had fought against the French and then the Japanese; now he wanted to keep France from controlling Vietnam again.

The French asked the United States to help them fight Ho and his forces (who were called *Vietminh*). They said Ho was a communist, and he was. He had gone to Russia and studied communism there. He believed in freedom for his country (which didn't necessarily mean freedom for the individuals inside the country).

Ho was an independent kind of communist. His goal was to free Vietnam from all outsiders. He needed help. When the Chinese communists began sending supplies

By the end of 1952, French casualties in Vietnam—dead, wounded, missing, captured—totaled more than 90,000. The French kept throwing more men, like these in 1953, against the Vietminh guerrillas. But the war was now unpopular at home and couldn't last much longer.

Confucius

Traditionally, the Vietnamese were followers of the ideas of Confucius. It is hard for most Western peoples to understand that Confucianism is not a religion in the same sense as Judaism or Christianity or Islam. Confucius said, "I am simply one who loves the past and is diligent in investigating it." Confucius was a teacher who taught *morality*, or the way to a good life. That good way is called Tao (DOW). He found its guidelines by studying the wisdom of the past. Americans and other Westerners do much thinking of the future: about invention and progress. Confucians try to perfect systems that have been tried over time. These are very different ways of looking at the world.

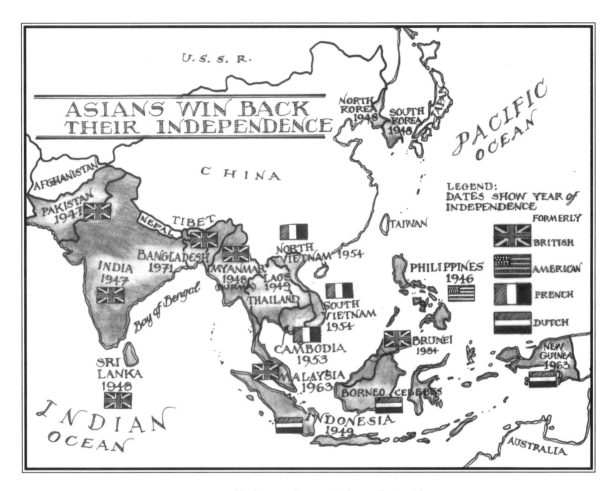

ASIANS WIN BACK THEIR INDEPENDENCE

U.S.S.R.

CHINA

NORTH KOREA 1948

SOUTH KOREA 1948

JAPAN

PACIFIC OCEAN

TAIWAN

LEGEND:
DATES SHOW YEAR OF INDEPENDENCE

FORMERLY

BRITISH

AMERICAN

FRENCH

DUTCH

AFGHANISTAN

PAKISTAN 1947

NEPAL

TIBET

BANGLADESH 1971

INDIA 1947

MYANMAR 1948 (BURMA)

NORTH VIETNAM 1954

LAOS 1949

THAILAND

SOUTH VIETNAM 1954

PHILIPPINES 1946

Bay of Bengal

CAMBODIA 1953

BRUNEI 1984

SRI LANKA 1948

MALAYSIA 1963

BORNEO

CELEBES

NEW GUINEA 1963

INDIAN OCEAN

INDONESIA 1949

AUSTRALIA

"**Unlike the** other countries of southeast Asia, Vietnam has always lived in the orbit of China," writes Frances Fitzgerald. For much of its early history, Vietnam was ruled by Chinese warlords. When Vietnam rejected Chinese rule, it kept much Chinese culture. Both China and Vietnam were Confucian countries. In the 20th century, China turned to communism; so did Vietnam.

to Ho he was happy to have their aid.

Perhaps if Ho Chi Minh had not been a communist we would have stayed out of the affair, but we soon sent military advisers to help the French. Then we gave France $10 million a year to fight Ho and the Vietminh. That was when Harry Truman was president.

By 1953 (when Eisenhower was president), we were spending $400 million a year to help the French in Vietnam, and before long almost twice that amount. It wasn't enough. The Vietminh kept winning. Vice President Richard Nixon and our top military leaders urged the president to send American bombers to Vietnam. Eisenhower refused.

The French prepared for a major battle. They were confident. They shouldn't have been. Their army was beaten and trapped, and most of the surviving soldiers died in a terrible jungle march. That was more than enough for the French people. They showed

French motorboats on patrol shortly before final defeat in 1954. At one point the French debated dropping an atom bomb on Vietnam.

in elections that they were ready to get out of southeast Asia. And, since France is a democracy, the people prevailed. A peace conference was held in Geneva, in Switzerland. There, Vietnam was divided into two sections: north and south. The division was meant to be temporary (most of the food-growing regions were in the south).

Ho Chi Minh was the leader of North Vietnam. He was now a great popular hero. After all, he had driven the French out of the country. South Vietnam's leaders were chosen by France. According to the Geneva agreement, elections were to be held within two years to reunite the country.

The elections were never held. The South Vietnamese leaders wouldn't allow them. They knew Ho Chi Minh would win. Soon there was civil war between the communist North and the pro-Western South. (Remember the situation in Korea? Was this the same?)

We were now out of Korea, and many of our nation's anti-communists believed that Vietnam was the place to take another stand against world communism. Those who wanted to send bombers and fighting troops were called *hawks*; those who didn't want to get more involved were called *doves*.

President Eisenhower said that if we let Vietnam become communist, it would be like watching a row of dominoes fall. The first domino would set off the others. Soon all of Asia would be communist. Eisenhower's advisers were hawks; they were urging him to fight. But none of them really understood southeast Asia. Hardly anyone in America did. We'd fired our China experts, who might have helped. Eisenhower was wary. "I'm convinced that no military victory is possible [there]," he said. (Remember that statement. You will read more about Vietnam.)

At Dienbienphu in 1954, the French were trapped in mountain jungles. The Vietminh brought up weapons and supplies along narrow roads, in columns of bicycles like these.

Hidden from sight behind their high hedges of bamboos, the villages stood like nuclei within their surrounding circle of rice fields. Within the village as within the nation the amount of arable land was absolutely inelastic. The population of the village remained stable, and so to accumulate wealth meant to deprive the rest of the community of land, to fatten while one's neighbor starved. Vietnam is no longer a closed economic system, but the idea remains with the Vietnamese that great wealth is antisocial, not a sign of success but a sign of selfishness.
—FRANCES FITZGERALD, *FIRE IN THE LAKE*

12 Separate but *Un*equal

"A lawyer's either a social engineer," said Howard University's law-school dean Charles Houston in 1935, "or he's a parasite on society."

Back near the end of the 19th century, Homer Plessy was arrested for sitting in a whites-only railroad car. Was it legal for the railroads to separate the races? What does the Constitution say?

The 14th Amendment says:

> *No State shall...abridge the privileges...of citizens of the United States...; nor deny to any person within its jurisdiction the equal protection of the laws.*

Abridge the privileges—that means take away or limit the rights of citizens. *Equal protection of the laws.* That seems clear. Does keeping people separate on a train abridge privileges? Does it deny anyone equal protection of the laws?

Some Americans weren't sure about that, and they looked to the Supreme Court for guidance. Finally, in 1896, the Supreme Court gave them an answer when it decided Homer Plessy's case. It was an answer that would cause a lot of people a lot of grief. Justice Henry Billings Brown wrote the decision for the majority of the justices. He said that the 14th Amendment called for

> *the absolute equality of the two races before the law, but...it could not have been intended to abolish distinctions based upon color, or to enforce social...equality, or a commingling of the two races.*

Do you understand that? The races were equal before the law, but laws could prevent them from mingling.

Justice John Marshall Harlan disagreed with the majority decision. Supreme Court justices often disagree with each other. The majority rules, but those who don't agree can write their *dissenting opinions.* In that famous 1896 decision, Justice Harlan wrote:

> *In view of the Constitution, in the eye of the law, there is...no superior,*

*dominant ruling class of citizens....*Our Constitution is color-blind, *and neither knows nor tolerates classes among citizens.*

But the majority opinion was the one that counted. The Supreme Court said that if facilities were equal they could be separate. The *Plessy* v. *Ferguson* decision made segregation legal in schools, restaurants, hotels, and public places in the southern states. Jim Crow had won the approval of the highest court. Separate but equal was the law.

Some people thought the Supreme Court had made a mistake. Some people thought the decision showed that the justices didn't understand the law of the Constitution. They agreed with Justice Harlan that "our Constitution is colorblind." One of those people was Charles

In 1963, more than 40 percent of Washington, D.C.'s families were African Americans living within sight of the Capitol—and below the poverty line.

Hamilton Houston. Houston graduated from Amherst College in 1915. He was an officer in World War I. After that, he went to Harvard Law School and got a law degree. Then he got still another college degree: a Ph.D. Even with all those degrees, Charlie Houston knew he had no chance of getting a job with a big law firm. His skin color would be held against him.

But Houston had no intention of working in a big law firm. He had studied law because he wanted to help his people. He believed that Jim Crow should be tried, sentenced, convicted, and hanged—and that the courts should do it. So Houston decided that he would become an expert in the law of the Constitution and then train other black lawyers to be experts too. And that is exactly what he did. He became dean of Howard University's law school. He was a very tough dean.

"He was so tough we used to call him 'Iron Shoes' and 'Cement Pants' and a few other names that don't bear repeating," said a student. "But he was a sweet man once you saw what he was up to."

What he was up to was making sure that his law students were as good as any lawyers anywhere.

"In all our classes," said another of Houston's students, "stress was placed on learning what our rights were under the Constitution...our rights as worded and regardless of how they had been interpreted to that time. Charlie's view was that we had to get the courts to change."

"He made it clear to all of us that when we were done we were expected to go out and do something with our lives," said Thurgood Marshall. Marshall was one of Charlie Houston's best students. He did something with his life—something important.

"Like an eating cancer," said Thurgood Marshall, segregation "destroys the morale of our citizens and disfigures our country throughout the world."

Jim Crow in the Far North

This restaurant in Juneau didn't just prevent Native Alaskans and Eskimos from eating within its precincts. It made sure they didn't work there, either.

Jim Crow, who stands for legal segregation, had a big residence in the South, but that miserable weasel (who kept a smile on his face to fool people) managed to settle in lots of other places too.

Alaska was one of them. Alaska? Yes. It was Jim who helped post signs there that said NO NATIVES ALLOWED on restaurants and hotels. And it was Jim who saw that schools were segregated. Whites sent their children to whites-only schools. Eskimos, Native Alaskans (Indians), and Aleuts went to other schools (although sometimes there were no other schools). Lots of non-white children just didn't get educated. They were illiterate, which means they couldn't read or write.

And then, because they couldn't read or write, the bigots said they were ignorant, "not civilized," and not fit to be with whites. You can see how frustrating this was for all decent people.

Back in 1905, an Aleut girl, a Miss Jones, wanted to go to the American public school in Sitka. Her father was white. But, since she went fishing with her Aleut grandmother, the judge said Miss Jones was "not civilized," and couldn't go to school.

Congress granted citizenship to all Native Americans in 1924, but that didn't integrate schools or end prejudice in Alaska. Then, during World War II, Americans everywhere began to look at prejudice with opening

eyes. It was hard to condemn Hitler's racist policies and accept racism at home. A reporter who visited Alaska in 1943 said that the social position of Indians and Eskimos was "equivalent to that of a Negro in Georgia or Mississippi."

In Nome, the Dream movie theater was segregated. Whites sat in one section, Native Alaskans and Eskimos sat in another. Alberta Schenck was an usher at the Dream. Her mother was Inupiaq, her father was white, and she hated the idea of segregation. But when she said something about it she was fired from her job. Alberta wrote a school essay about her feelings. She said:

I believe we Americans and also our Allies are fighting for the purpose of freedom. I myself am part Eskimo and Irish and so are many others. I only truthfully know that I am one of God's children regardless of race, color, or creed....What has hurt us constantly is that

Ernest Gruening, governor of Alaska from 1939 to 1953; he backed desegregation.

we are not able to go to a public theater and sit where we wish, but yet we pay the same price as anyone else and our money is gladly received.

That, said Alberta, was "following the steps of Hitlerism."

A few weeks later, Alberta had a date with a white sergeant from a nearby army base. They went to the movies. They sat in the whites-only section. The manager ordered her to move. "Get over there with the Eskimos!" he yelled. "Don't move," said the sergeant. "You're my guest." The manager called the Nome chief of police. The chief grabbed Alberta, pulled her down the aisle, and took her to jail. Alberta Schenck spent the night in the Nome city jail.

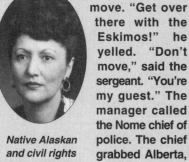

Native Alaskan and civil rights fighter Elizabeth Peratrovich.

Then she wrote Governor Ernest Gruening. She said, "My father was a soldier in World War I. I have two brothers in the army in this war." And she told him what had happened to her at the Dream theater.

The governor (a New Deal appointee) was furious. He said, "I consider it un-American...I deem it contrary to the spirit of our country and directly in conflict with the issues on which this great war is being fought."

Some Alaskans had been working hard to see segregation outlawed. Elizabeth Wanamaker Peratrovich and her husband, Roy, were two of those people. They moved to Juneau in the early 1940s, and found they could not buy a house in the part of town where they wanted to live. They were turned away because they were Alaskan natives. Elizabeth, a Tlingit, was president of the Alaska Native Sisterhood. She was determined to get an anti-discrimination act passed. It would make segregation illegal. The bill was defeated when it first came before the legislature in 1943. In 1944, it was defeated again. That was the year that Alberta Schenck spent a night in jail.

Allen Shattuck, the senator with 5,000 years of civilization behind him.

Governor Gruening wrote to Alberta and said he would work to see that the anti-discrimination bill was passed in the next legislative session. "If it becomes law, you may be certain that the unpleasant experience which has been yours will not happen again to anyone in Alaska."

But some people were opposed to the bill. Senator Allen Shattuck of Juneau spoke out in the Senate and said:

Far from being brought closer together, which will re-sult from this bill, the races should be kept apart. Who are these people, barely out of savagery, who want to associate with us whites with 5,000 years of recorded civilization behind us?

Elizabeth Peratrovich was sitting in the Senate gallery. She rose and, in firm tones, answered the senator:

I would not have expected that I, who am barely out of savagery, would have to remind the gentleman with 5,000 years of recorded civilization behind him of the Bill of Rights.

The bill passed the Senate and the governor signed it on February 16, 1945. In Alaska today, that date is celebrated as Elizabeth Peratrovich Day.

Governor Gruening signs Alaska's Anti-Discrimination Act in 1945; Elizabeth Peratrovich is on his right.

13 Linda Brown— and Others

In 1954, segregation was legal if the facilities provided to blacks and whites were equal. This one-room North Carolina schoolhouse (for seven classes) contains a "library," "running water," and "central heating." See if you can find those things in the picture.

Linda Carol Brown—who was seven years old and lived in Topeka, Kansas —had to walk across railroad tracks and take an old bus to get to school, though there was a better school five blocks from her house. Linda couldn't go to that school because she was black and the schools in Topeka were segregated. Linda's father, the Reverend Oliver Brown, didn't think that was right. He went to court to try to do something about it. Their case became known as *Brown* v. *Board of Education.*

South Carolina's Clarendon County spent $43 a year on each of its black students. It spent $179 a year on each white student. The white children all had school desks; in two of the black schools there were no desks at all. Harry and Liza Briggs and 20 other black parents sued the Clarendon County school board. They wanted equal funding for the black schools. They sued in the name of 10-year-old Harry Briggs, Jr., and 66 other children. Right away, Liza Briggs was fired from her job. So were most of the other adults who signed the lawsuit that was titled *Briggs* v. *Clarendon County.*

Barbara Rose Johns, a junior at Moton High School in Farmville, Virginia, was angry about conditions in her school: it had been built for 200 students, but held 450. There was no cafeteria and no gym. The highest-paid teachers at Moton received less than the lowest-paid teachers at Farmville's white schools. A committee of black parents had petitioned the county for a new school and had been turned down.

Racism wasn't (and isn't) just a problem of blacks in the South. It is a human problem, and it is found across the land (and across the world too). Racism is the irrational hatred of those who are different from you. Almost every minority group has, at one time or another, suffered the pain of irrational hatred.

Linda Brown in 1951, the year her father sued Topeka's Board of Education to let her go to her local school.

Johns decided to act. She had a friend telephone the school principal and tell him he was needed at the bus terminal—at once. Then she called a meeting of all the students. She told the teachers they were planning a surprise event.

She was right; they were surprised. Barbara Johns talked the students at Moton into going on strike for a better school. They walked out of their classes. A member of the National Association for the Advancement of Colored People (NAACP) came to Farmville. He intended to tell the students that Farmville was not the place to fight segregation. But he was so impressed with their determination that he helped 117 Moton High School students sue the state of Virginia. They demanded that the state abolish segregated schools. Their case was called *Davis* v. *County School Board of Prince Edward County* because the first of the students listed was 14-year-old Dorothy E. Davis.

Each of those three cases was defeated in court, but that didn't stop the plaintiffs (those who were suing). They appealed the cases. They appealed them all the way to the United States Supreme Court. There they were grouped with two other cases dealing with school segregation: one from Delaware, and one from Washington, D.C. Together the five suits were called by the name of the first of them: *Brown* v. *Board of Education.*

That case would directly affect all the schools in the 21 states with segregated schools. It would indirectly affect almost every school in the United States. *Brown* v. *Board of Education* was to become one of the most important cases ever brought before the Supreme Court.

Supreme Court cases are not handled like the cases you see on television. The people involved—the schoolchildren, in this case—don't come before the court. There are no witnesses. Lawyers do all the talking. They often spend months—or years—preparing their cases. Then they present their argument to the nine justices. The Supreme Court justices usually ask questions. Sometimes those questions can be answered at once. Sometimes the lawyers have to come back and re-argue the case.

Anyone can attend a Supreme Court session if he or she is willing to stand in line. The court is in Washington,

What does *appealing* a law case mean? Our legal system begins with city and county courts, goes on to state courts and then to federal courts, and, finally, to the Supreme Court. Suppose you go to court and you don't think your trial was fair. You can appeal your case to a higher court. The higher court may reverse the lower court's decision. If it doesn't, you may, in some cases, appeal the case still further, and further, until finally you get to the Supreme Court.

Thurgood Marshall (center) with the lawyers working on the *Brown* case. They practiced their arguments on the students and faculty of Howard Law School.

67

In 1963—another landmark year for civil rights, as you'll discover—Ben Shahn made this painting of the 1954 Supreme Court justices who handed down the historic *Brown* ruling.

"Sometimes I think I'm happiest when I've forgotten myself for a long, long time," a mere eight-year-old black child told me...as she struggled in the face of a hostile mob to enter an all-white southern school....But she persisted, she endured, and she always and thereafter called that time her "big chance." She had stumbled the hard way upon wisdom, upon grace, upon a kind of release based upon moral purpose; and maybe many of us, so much better off in our lives, may still be waiting for *our* "big chance."

—Robert Coles, *Happiness*

D.C., right near the U.S. Capitol (which is where Congress meets). On December 9, 1952, all the seats were filled in the Supreme Court chamber and 400 people were turned away. That is unusual, but this was an unusual day. The court was ready to consider *Brown* v. *Board of Education.*

The NAACP was representing the children. Charlie Houston was dead, but his star pupil, Thurgood Marshall, the great-grandson of a slave, argued their case. Marshall, a hard worker, was a meticulous lawyer with a good sense of humor. He had argued 15 cases before the Supreme Court—and won 13.

The lawyer Marshall faced was John W. Davis. Some people said that Davis was the best lawyer in America. He had argued more cases before the Supreme Court than any living attorney. Davis had run for president against Calvin Coolidge, and everyone seemed to like him (though he lost the election). King George V of England said John Davis was "the most perfect gentleman" he had ever met. Even Thurgood Marshall liked Davis; they often ate lunch together.

Supreme Court justices do not decide questions of right and wrong. That can sometimes be a matter of opinion. We live in a society based on law. The Constitution is our highest law. The job of the justices is to decide the meaning of the Constitution. Does the Constitution permit segregation? Or does segregation break the rules laid out in the Constitution? That was the question the justices had to decide.

Marshall and the NAACP lawyers presented two arguments: first they argued that the 14th Amendment—which says *No State shall...abridge the privileges...of citizens of the United States...; nor deny to any person within its jurisdiction the equal protection of the laws*—made the doctrine of "separate but equal" unconstitutional. Then they argued that segregated schools can never be truly equal—separating people, of itself, makes them feel unequal and inferior. The *Plessy* decision was wrong, they said.

John Davis looked at the 14th Amendment and the rest of the Constitution. He said that nothing in it prevented separation, as long as equal facilities were provided. Each state, said Davis, has the right to make its

In some places integration of schools was very slow; in some places it didn't seem to happen at all. In others, like Louisville, Kentucky, where these children attended school, enlightened superintendents made the change-over work quite smoothly.

Here is part of what he read:

own decisions on social matters such as segregation. He believed that the *Plessy* decision was right.

This was a very difficult case. The justices asked questions. They took their time. A year passed. It looked as if the court might be split, with some justices saying segregated schools were unconstitutional and some saying they were not. This issue was dividing the country. If the court were to split, it would make those divisions worse. Then something unexpected happened: the Supreme Court's chief justice died. President Dwight Eisenhower named California's former governor Earl Warren as the new chief justice. Warren was a mild-mannered man who was not expected to be a dynamic chief justice. But a few people who knew him well understood that he had a gift for leadership. They also knew that he had a strong moral sense: he believed in justice and fairness.

Finally, the waiting was over. On May 17, 1954, Earl Warren read the decision in *Brown* v. *Board of Education.*

> *It is doubtful that any child may reasonably be expected to succeed in life if he is denied the opportunity of an education. Such an opportunity…is a right which must be available to all on equal terms….Does segregation of children in public schools solely on the basis of race… deprive children of the minority group of equal educational opportunities? We believe that it does….We conclude, unanimously, that in the field of public education the doctrine of "separate but equal" has no place. Separate educational facilities are inherently unequal.*

UNANIMOUSLY! The new chief justice had convinced all the justices that, because of the importance of this decision, it should be unanimous (which means they should all agree). It was, as the *Washington Post* said the next day in an editorial, "a new birth of freedom." *Plessy* v. *Ferguson,* a case about a railroad car, had made segregation a fact in al-

Speaking Out

Thirteen-year-old Mary Beth Tinker came to school wearing a black armband to protest the Vietnam War. The principal told her to remove it. Mary Beth thought the armband was a form of speech—symbolic speech. She refused to take it off, was suspended from school, and went to court. Her case, *Tinker* v. *Des Moines Independent School District,* went to the Supreme Court. Here is what the court said in its 1969 decision:

School officials do not possess absolute authority over their students. Students in school as well as out of school are "persons" under the Constitution. They are possessed of fundamental rights which the State must respect, just as they…must respect their obligations to the State.

For more on the *Tinker* case, and others involving young people, read Nat Hentoff's *American Heroes: In and Out of School.*

Chief Justice Earl Warren.

69

Third-graders in Wilmington, Delaware, in 1961, seven years after *Brown* v. *Board of Education* was handed down.

How did Brown *v.* Board of Education *affect schools in your district?*

most all phases of daily life in the South. *Brown* v. *Board of Education*, a case about schoolchildren, would provide a way to attack segregation—and not just in the classroom.

But the battle wasn't over. Laws have to be enforced, and some people were determined not to enforce this one. Virginia's Prince Edward County closed all its public schools—for *five years*—rather than integrate the schools. White children were educated in "private" white academies funded with tax money (paid by white and black taxpayers). Black children were denied any schooling at all. Prince Edward County wasn't alone in its mean-spiritedness. Most southern communities refused to integrate schools. In Norfolk, Virginia, all public schools were closed for a year. Most children—black and white—didn't have any schools to go to.

It was a difficult time for moderate southern whites. They had always lived with segregation. Strong voices were shouting that the southern world they knew and loved would end if they agreed to integrate their schools. (It was the same message that had been used to defend slavery 100 years earlier.) Those who spoke out against segregation often lost their jobs and friends. Some white people were scared.

Where were the voices of reason? The moderate southern leaders seemed to have gone into hiding. But, remember, these were conforming times. All over the nation, people were keeping silent while others were abused.

In a few areas—especially in the states bordering the North—integration proceeded, usually without incident. White children had no trouble going to school with black children. It was the adults who were creating problems. They were dragging out all the old, tired arguments.

In some places, when black children marched into integrated schools, grownups insulted them, or threw rocks. Because of that new medium—television—everyone, all over the world, could see the rocks and the taunting faces. Decent folks hid their heads in shame. *Brown* v. *Board of Education* may have been a new birth of freedom, but the baby was having a hard time breathing on its own.

Nettie Hunt explains to her daughter Nikie the meaning of the Supreme Court's *Brown* decision, about which the *New York Times* said: *The highest court in the land, the guardian of our national conscience, has reaffirmed its faith and the underlying American faith in the equality of all men and all children before the law.*

Getting Integrated: Two Southern Experiences

In the tiny mountain town of Piedmont, West Virginia, school commissioners integrated schools soon after the Supreme Court's decision. Henry Louis Gates, Jr., who is black, describes how that affected his life:

Less than four years after my birth, something happened that would indelibly mark me and my peers for life—something that would open up another world to us, a world our parents could never have known. *Brown* v. *Board* was decided in 1954.

I entered the Davis Free Elementary School in 1956, just one year after it was integrated. There are many places where the integration of the schools lagged behind that of other social institutions. The opposite was true of Piedmont. What made the Supreme Court decision so determining for us was that school was for many years after 1955 virtually the only integrated arena in Piedmont....

We were the pioneers, people my age, in cross-race relations, able to get to know each other across cultures and classes in a way that was unthinkable in our parents' generation. Honest hatreds, genuine friendships, rivalries bred from contiguity rather than from the imagination. Love and competition. In school, I had been raised with white kids, from first grade. To speak to white people was just to speak. Period....

Only later did I come to realize that for many colored people in Piedmont...integration was experienced as a loss. The warmth and nurturance of the womblike colored world was slowly and inevitably disappearing, in a process that really began on the day they closed the door for the last time at Howard School, back in 1956.

Kathryn Harris Morton, who is white, was a schoolgirl in 1958 in Norfolk, Virginia, when her school, like many in Virginia, was closed to all students to prevent integration. This is what she says of that time of "massive resistance":

I was one of the students locked out of classes and one of the 17 children who together successfully sued the governor, the attorney general, and Norfolk's school superintendent, thus ending the lockout.

I remember the Norfolk Committee for the Public Schools...the public rally...the newspaper coverage, the petition we took from door to door asking people to say they favored public education. I remember neighbors saying they would be afraid for their jobs if they put their signatures on such a request....I remember the resounding public silence of Norfolk's business leaders during all the months of the school closing.

November 1958 we had our day before the special three-judge federal court. Our lawyers charged that we were being denied due process and equal protection under the law. There had been a struggle to find a lawyer willing to risk his career by taking so unpopular a case. I remember the threatening phone calls late at night, the eggs thrown at our house, our Halloween pumpkin blown up on our porch, our picture in *Life* magazine. Neighbor children said, "Why do you bother? Don't you know they'll take care of it?"

And when [the judges] came down with their decision on a point of law, ending massive resistance and opening the schools, the same children said, "See, they took care of it." I remember laughing the next morning to see that the business "leaders" of Norfolk finally had something to say. Now that the battle was decided, they took a full-page ad supporting public education. Leading from the rear is safer than risking those midnight phone threats.

Now, as the story is retold and retold, the "leaders" of Norfolk are airbrushed into the picture of what happened that fall and winter. Doubtless many of those high-minded men wanted to do something when their actions would have made a difference....

Still, I hope the next reporter or historian who plans to write again about the school closing will go look at the [newspaper] coverage at the time. Readers should not be misled into thinking when a public problem arises that individuals don't need to fret themselves, under the delusion that "the leaders" can be counted on to take care of it.

14 MLKs, Senior and Junior

Martin L. King, Sr., born in Stockbridge, Georgia, became pastor of Ebenezer Baptist Church in Atlanta.

Mike King was a Georgia sharecropper's son who was teased when he went to school because he smelled of the barnyard. "I may smell like a mule," he said, "but I don't think like one." He could handle the teasing—but he couldn't handle his father. The man drank too much and was violent when he was drunk. Mike's mother had saved some money, so she bought him an old car—a Model T Ford—and he headed off to the big city: Atlanta. As he drove away from home, he passed a two-story brick house that belonged to a banker. "I'm going to have a house like that," he said, "and I'm going to be a bank director, too."

King was muscular, energetic, and ambitious, with a zest that attracted people. In Atlanta he worked on the railroad and discovered that what he really wanted to do was preach. So he did that, on Sundays, in small Baptist country churches. Then he learned of a girl, Alberta Williams, who was the daughter of the leading Negro preacher in Atlanta. She was a stu-

What's in a Name?

*N*egro, black, colored, *Afro-American,* and *African American* are all words used to describe those who are descended, in whole or in part, from people of African origin. Words, like other things, have fashions. *Negro* was the preferred term for many generations. Today, *African American*—or *black*—is most people's choice.

Martin Luther King, Jr., met Coretta Scott in Boston, where she was studying singing. This photo shows them with three of their four children in 1963, the day after an Atlanta school refused to admit their son Martin Luther III (left).

dent at Spelman College and a talented organist. Before they even met, he decided he would marry her.

In the segregated South, the church was the center of the black world. It was the place where you took your troubles and your heart and found friends and support in an often unkind world. The minister was apt to be the most respected and best-educated man in the black community.

There was no way a rough country boy could marry the daughter of the Reverend A. D. Williams. Why, both her mother and her father had college degrees. Young King couldn't even speak properly. Well, Mike King decided, if it was necessary, he would get an education. So he went to a public school and said he wanted to learn. They tested him and found he was barely ready for fifth-grade work. King was 20. They brought a big desk into the fifth grade and he sat there with the 10-year-olds and did his lessons and worked at night. A few years later he was finished with high school, but that wasn't enough. To be accepted in the Williams family he needed a college degree. So he went to Morehouse College—where the Reverend Williams had studied—took the entrance tests, failed them, and was turned away. But there was no stopping Michael Luther King. He marched into the college president's office, right past an exasperated secretary, told the president he wanted to go to college and that he would work hard and do well, which is exactly what he did. He got a college degree; and he got the girl he wanted; and, eventually, he got his father-in-law's church, and a two-story brick house—and he even became a bank director.

But what he was proudest of was his family. He and Alberta, who was church organist, took their three children everywhere and beamed with pride at their accomplishments. His older son, named for him and called M.L., was a small, wiry, athletic kid who loved to play ball and had lots of friends. When M.L. was five, his father changed their names. Each became Martin Luther King, after the German priest who had founded Protestantism.

Martin Sr. and Alberta made sure their son had a good education. They sent him to a laboratory school at Atlanta University, to segregated Booker T. Washington High School, and, at age 15, to Morehouse College. Martin intended to be a doctor. During the summers—they were World War II years—he went off to Connecticut and picked tobacco. There he changed his mind: he would become a minister. His father told the church members that young Martin had been "called by God to the pulpit." But Martin's friends joked that it was the hot sun of the tobacco field that had something to do with his decision.

Martin Luther King, Jr., chose Crozer Seminary, in Pennsylvania, to study theology—about religion—and he chose well. That small, elite

After he grew up and became famous, Martin Luther King, Jr., wasn't often seen laughing in public. But he had a great sense of humor and loved to tell jokes with friends.

Crozer Seminary no longer exists.

King with a portrait of Gandhi, who, he said, taught him that "there is more power in socially organized masses on the march than there is in guns in the hands of a few desperate men."

Captain Charles

Boycott was a 19th-century Irish land agent and owner whose tenants got mad when they thought his rents were too high in a time of poor crops. They refused to cooperate with him. Without workers, his land, crops, and animals were useless. Local merchants wouldn't supply him. Eventually Boycott was forced off his land, but his name passed into the English language. To *boycott* something is to refuse to use it or have anything to do with it.

school had white and black students from North and South, some Asian-Americans, several American Indians, and students from other lands. It was an astounding mix—unique in its time. For a privileged boy from a protective family, living with all those people was an education in itself.

Martin Luther King, Jr., who had been a cutup at Morehouse, became the valedictorian—the top student—in his class at Crozer. Books were piled high on his bedroom floor; sometimes he read all night. There he discovered that he had a passion for words and ideas and a talent for public speaking.

At Morehouse, King had read a book by Henry David Thoreau called *Civil Disobedience*. Thoreau believed in the power of nonviolence. He believed in the power of even "one honest man" to create great change in the world.

At Crozer, King learned about India's great leader Mohandas (Mahatma) Gandhi, and that Gandhi had been inspired by Thoreau. Gandhi, a small, skinny lawyer with a squeaky voice, was the honest man Thoreau believed in. He led millions of people in nonviolent boycotts and marches to protest British rule in India. When British soldiers taunted and beat and jailed Gandhi and his followers, they didn't fight back with fists or guns; they just kept peacefully marching and protesting. Gandhi's reasoned courage and calm dignity turned away the guns and cannons of a mighty empire. Gandhi showed the world the power of goodness and right action. India became free. Martin Luther King was fascinated to discover that Gandhi had been full of rage when he was young and had learned to control his anger. King was often angry. Could he teach himself self-control? Could he teach others?

Martin Luther King, Jr.'s new learning seemed to expand his ideas on Christianity and Christian love. In his father's world, Christianity was simple and sure; at Crozer, King found a Christianity that was questioning.

He still wasn't finished with school. After Crozer he went to Boston University for more study and reading and for a Ph.D. (that made him the Reverend *Dr.* Martin Luther King, Jr.). His professors wanted him to become one of them: a teacher and a scholar. But Martin wanted to be a preacher. He had an idea that a minister could do things to make the world better. He wanted to fight injustice. He wanted to lead his people—the black people—because he thought they had a message for all people. Segregation and racial hatred were wrong. Injustice and unfairness were wrong. Social cruelties and meanness hurt everyone (like a worm that turns a good apple rotten).

The more Martin Luther King, Jr., thought about Thoreau, and Gandhi, and about Christian ideas on loving your enemies, the more he began to believe in the power of peaceful protest. When he thought about America's founding idea, that "all men are created equal," he wondered if nonviolent action could be used to actually bring that equality to all people. Could nonviolence overcome the evil of segregation? He knew it wouldn't be easy.

How *do* you face evil?

You can turn away from it, which is the easiest thing to do.

You can fight it with weapons or fists—which is harder, and may hurt or kill people.

The hardest way of all is nonviolence. It means standing up to evil without weapons. It means taking punches and not returning them. Now that takes courage.

"I went to bed many nights scared to death by threats against myself and my family," said Dr. King to his Montgomery congregation. "Then in my kitchen...I heard a voice say, 'Preach the gospel, stand up for the truth, stand up for righteousness.' Since that morning I can stand up without fear."

Martin had a lot to think about. He wasn't sure how he would lead his people—he just knew that was what he wanted to do. In the meantime, his father was getting impatient. He wanted his son as assistant pastor at his fine big church in Atlanta, Georgia. But Martin Jr. decided he would start his career at a small church in a quiet city. He had no idea that an explosion—called the *civil rights movement*—was about to begin in that quiet city.

It was Montgomery, Alabama, and soon everyone in America would know about it.

To **taunt** is to jeer, tease, and goad with words.

15 Rosa Parks Was Tired

"I handle and work on clothing that white people wear," said Rosa Parks. "This is what I wanted to know: when and how would we ever determine our rights as human beings?"

Rosa Parks, who worked as a tailor's assistant in a department store in Montgomery, Alabama, was a small, soft-voiced 43-year-old woman who wore rimless glasses and pulled her brown hair back in a bun. Parks had been secretary of the Montgomery chapter of the NAACP, so she was well known to Montgomery's black leaders. She was also well respected. Rosa Parks was refined and reliable.

But on the evening of the first day of December in 1955, Mrs. Parks was mostly just plain tired. She had put in a full day at her job. She didn't feel well, and her neck and back hurt. She got on a bus and headed home.

In 1955, buses in all the southern states were segregated. Laws said that the seats in the front were for whites, those in the back for blacks. Parks sat down in the section for blacks. Then, when all the seats filled up, the driver asked Parks to stand and give her seat to a white man (that was customary in Jim Crow Alabama). Rosa Parks wouldn't budge. She knew she might get in trouble, she might even go to jail, but suddenly she found herself filled with determination. She stayed in her seat.

The bus driver was filled with rage. He called the police. Rosa Parks was soon arrested and on her way to jail. She knew that blacks were beaten and abused in Montgomery's jail. It didn't seem to matter to her. She was tired of riding on segregated buses. She was tired of being pushed around. She was even ready to go to jail.

When the ministers and black citizens of Montgomery heard of her arrest, they were stunned. Of all people—mild-mannered, dignified Mrs. Parks in jail? E. D. Nixon, who had been president of the local NAACP

"The Rosa Parks incident was completely spontaneous," said Parks's attorney. "You have to know the history of how people have been treated on the buses. Buses have been the sorest spot in Montgomery."

chapter, raised bond money to get her out of jail. But she would have to go on trial for breaking the law—the segregation law.

Rosa Parks knew that some of Montgomery's black leaders (members of the NAACP, and a group of professional women) were trying to find a way to do something about segregation on the city's buses. E. D. Nixon asked her if the NAACP could use her case to fight segregation. They both knew that might put her life in danger. Blacks who stood up for their rights were sometimes lynched (hanged). Parks talked to her husband and her mother. Then she thought a while and said quietly, "I'll go along with you, Mr. Nixon."

As soon as Jo Ann Robinson heard that Rosa Parks had been arrested, she began organizing a boycott of the buses. Robinson felt humiliated and angry each time she had to sit in the back of a bus. She had longed to do something about Jim Crow buses. Now she was ready to act. She decided to ask all Montgomery's blacks to stay off the buses for one whole day as a protest. Robinson and some friends stayed up most of the night (it was a Thursday). They printed leaflets—35,000 of them—telling the black community to keep off the buses the next Monday, the day of Rosa Parks's trial.

Montgomery's leading Negro ministers agreed to support the one-day boycott. In their sermons on Sunday they urged everyone to stay off the buses on Monday. They knew that wouldn't be easy. Those who rode the buses were mostly the poorer citizens. They were people who needed to get to work. Some were elderly. It was December, and cold. Some

Artist Marshall Rumbaugh commemorated Rosa Parks's determination in this painted wood sculpture. "I feel sure we will win in the end," said Parks. "It might take plenty of sacrifice, but I don't think we'll ever again take such humiliating treatment as dished out by bus drivers in this city."

When E. D. Nixon went to get Rosa Parks out of jail, two white friends went with him. They were Virginia and Clifford Durr. Durr, a lawyer, helped the boycott movement, and because of that lost many white clients. The Durrs were not the only whites who sided with the marchers. White minister Robert Graetz's house was bombed. Others lost their businesses or had to put up with insults and threats.

could find rides, but many would have to walk miles. And they all feared white violence. It was customary to intimidate blacks who tried to stand up for their rights. It was fear that made segregation work.

But something unexpected happened in Montgomery. Like Rosa Parks, most black people no longer seemed afraid. They had had enough. They stayed off the buses on Monday. And also on Tuesday. And then all week. And all month. And on and on, in rain and cold and sleet and through the heat of summer. They stayed off the buses. They

Right, Rosa Parks is fingerprinted after her arrest. Four days later, she was convicted of violating Montgomery's segregation laws and fined $14. On the same day, the bus boycott began, and all over town black citizens organized carpools to get themselves to work (below).

shared rides; they worked out elaborate carpools; they walked. Houses were burned, churches were bombed, and shots were fired, but Montgomery's black people stayed off the buses. The jails filled with people whose only crime was riding in a carpool; still the boycott continued.

Montgomery's black citizens had always thought of themselves as ordinary folk, but they were proof that ordinary people can do extraordinary things. They were lucky: what they were doing was important, and they knew it. They were doubly lucky: they had remarkable leadership. The community had several strong leaders, but one was outstanding.

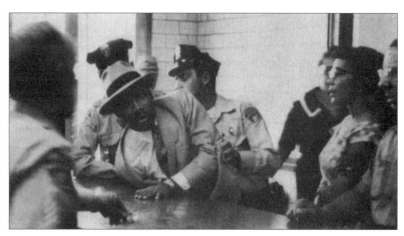

That leader was a 26-year-old minister, newly arrived in Montgomery, with a pretty, musically gifted wife named Coretta and a two-week-old daughter. He was Martin Luther King, Jr.

When King was asked to lead the boycott, he accepted. He was a beginner—as a minister and a leader—and he didn't know what he would have to do. But he and others found out that when it came to strength and vision and courage, Martin Luther King, Jr., had what was needed.

And so a great leader, a just cause, and an inspiring idea came together at the same time, and made history.

The cause began with Rosa Parks. It was the cause of fairness. Segregation is unfair. Segregation is humiliating. Segregation is wrong.

The idea was nonviolence. King had been considering that idea, and he wasn't the only one. A number of others—black and white—believed Gandhi's methods would work in America. It was King who took the idea, spoke it in powerful words, and inspired others to act on it.

"We are not here advocating violence," he said in Montgomery. "The only weapon that we have...is the weapon of protest...[and] the great glory of American democracy is the right to protest for right." And so, while the segregationists screamed bad words and kicked cars and set off bombs, Montgomery's black community protested with calm, unflinching courage. They didn't scream back. They maintained their dignity. Newspaper reporters began to come to Montgomery to see what was happening. Television crews came too. Soon people around the nation, and in other nations as well, were watching the people of Montgomery marching. They marched to work; they marched to well-organized

Montgomery, 1955: Dr. and Mrs. King are arrested for organizing the bus boycott. "Where segregation exists we must be willing to rise up en masse and protest courageously against it," said Martin Luther King. "I realize that this type of courage means suffering and sacrifice. It might mean going to jail. If such is the case we must honorably fill up the jails of the South."

The amazing thing about our movement is that it is a protest of the people. It is not a one-man show. It is not the preachers' show. It's the people. The masses of this town, who are tired of being trampled on, are responsible.
—Jo Ann Robinson, 1955

Ralph Abernathy (left) greets Dr. King just after his 1956 conviction for leading the boycott. Nine months later, the boycott over, they boarded the bus together to sit up front. "We are glad to have you this morning," said the bus driver.

They'll find that all they've won in their year of praying and boycotting is the same lousy service I've been getting every day.

—WHITE MONTGOMERY BUS RIDER, DECEMBER 1956

The boycott over, Rosa Parks rides the bus again—at the front. Why wasn't she riding on that first desegregated bus with the male leaders? Could it be that women weren't taken seriously by most men—white or black? More on that later.

carpool centers. When they were arrested, they marched to jail. TV watchers also saw and heard the screamers and rock throwers. And they listened to Martin Luther King, Jr.'s eloquent words:

> There are those who would try to make of this a hate campaign. This is not a war between the white and the Negro but a conflict between justice and injustice. If we are arrested every day, if we are exploited every day, if we are trampled over every day, don't ever let anyone pull you so low as to hate them. We must use the weapon of love.

The weapon of love won the battle. Thirteen months after Rosa Parks's arrest, the Supreme Court ruled that segregation on Alabama buses was unconstitutional. The boycott was ended! Martin Luther King, Jr., E. D. Nixon, Ralph Abernathy (a black minister and boycott leader), and George Smiley (a Texas-born white minister) rode the first integrated bus—and they all sat up front together.

The people of Montgomery not only changed their world, they changed their times.

16 Three Boys and Six Girls

Some white children in segregated schools wanted to try integration. "They don't want you to think for yourself," said one Central High student. "Let us try it. Make the parents go home." But others, like those above, needed to learn about fairness—and about spelling, too.

The fight to see that all Americans—black, white, Hispanic, Asian, female—would be treated fairly was called the *civil rights movement*. Some of its most important battles were fought by school students.

After the Supreme Court announced its decision in *Brown* v. *Board of Education* in 1954, the court said integration should take place with "all deliberate speed." What does that mean? The southern states decided that it meant with the speed of a snail. So, in the Deep South in 1957, there were still no classrooms where black boys and girls and white boys and girls sat together. Then a federal judge ordered schools in Little Rock, Arkansas, integrated.

Little Rock's Central High School was built in 1928. Some people, then, called it the finest public high school in the nation. Twenty-nine years later, it was still a good school. It had generous playing fields, modern facilities, and more than 2,000 students. But not one black child had ever gone to Central High. In Arkansas, as in all the southern states, laws said that blacks could not go to public schools with whites.

Melba Pattillo wanted to go to Central High. "They had more equipment, they had five floors of opportunities. I understood education before I understood anything else. From the time I was two, my mother said, 'You will go to college. Education is your key to survival,' and I

You just realize that survival is day to day and you start to grasp the depth of the human spirit, and you start to understand your own ability to cope no matter what. That is the greatest lesson I ever learned.

—MELBA PATTILLO

81

Nine Brave Kids

In 1987, 30 years after they entered Central High, the Little Rock nine came together for a reunion. Elizabeth Eckford, now a social worker, was the only one who had stayed in Little Rock. Thelma Mothershed was a teacher in Illinois, Terry Roberts a professor at UCLA, Minnijean Brown a writer and mother of six, Jeff Thomas a Defense Department accountant in California, Ernie Green a vice president of a New York investment firm, Carlotta Walls a Denver realtor, Gloria Ray a magazine publisher living in the Netherlands, and Melba Pattillo a communications consultant and author living in San Francisco. Do you think they might have been strengthened by their struggle?

Gloria Ray   Terrence Roberts   Melba Patillo

Elizabeth Eckford   Ernest Green   Minnijean Brown

Jefferson Thomas   Carlotta Walls   Thelma Mothershed

The *Chicago Defender* said: "The Supreme Court ruling would have been meaningless had these Negro boys and girls failed to follow the course mapped out for them by the law....They should be applauded by all of us."

When Melba Pattillo got to her English class on her first day of school, "One boy jumped up to his feet and began to talk. He told the others to walk out with him because a 'nigger' was in their class. The teacher told him to leave the room." Melba went on, "The boy started for the door and shouted: 'Who's going with me?' No one did. So he said in disgust, 'Chicken!' and left. I had a real nice day."

understood that." Otherwise, Melba said, she had no "overwhelming desire to go to this school and integrate this school and change history."

But 15-year-old Melba would change history. She was one of nine black children to integrate Central High. At first, she didn't expect problems. Neither did most other people. Little Rock's citizens thought their city had good race relations. But some people in Little Rock decided to fight integration. They used threats, rocks, and nasty words.

Others, who might have shown some courage, kept quiet. Arkansas's governor, Orval Faubus, announced that he would call in the National Guard. Most people thought the guardsmen would protect the black students, but Faubus meant to use them to keep the nine out of Central High. He knew that aiding integration would make him lose white votes. (And blacks weren't able to vote, so they didn't matter to him.)

Later, Melba remembered:

The first day I was able to enter Central High School, what I felt inside was terrible, wrenching, awful fear. On the car radio I could hear that there was a mob. I knew what a mob meant and I knew that the sounds that came from the crowd were very angry. So we entered the side

*of the building, very, very fast.
Even as we entered there were people running after us, people trip-
ping other people....There has never
been in my life any stark terror or
any fear akin to that.*

Melba Pattillo had reason to be afraid.
The mob was threatening to hurt her and
the other black students. Slim, shy Elizabeth Eckford faced the mob alone. Elizabeth
was wearing a starched new black-and-
white cotton dress for her first day at
school. She had not gotten the message
that the black students were to enter school
together. So she was by herself, at the opposite end of the building from the others.
Elizabeth must have been scared, but she
held her head high and tried to walk up to
the school door. The guardsmen stared at
her. Adults screamed awful words. A woman spat. Some boys threatened to lynch
her. Elizabeth ran back to the curb, and a
New York Times reporter put his arm around
her. "Don't let them see you cry," he whispered. A white woman (who was on her
side) took Elizabeth home.

Three black reporters weren't as lucky.
They were beaten by enraged whites. One
of them, Alex Wilson, was a former marine
and more than six feet tall. He was hit with
a brick and "went down like a tree."

In the nation's capital, President Eisenhower said he didn't want to take sides. He
believed in persuasion. But there was no
persuading the lawbreakers who stood outside Central High that day and the next.

Finally, the president acted. "Mob rule
cannot be allowed to override the decisions
of our courts," he said. Reluctantly, he ordered federal troops sent to Little Rock.

"The troops were wonderful," said Melba

Elizabeth Eckford goes to school. "Around the massive brick schoolhouse," wrote Daisy Bates, president of the Arkansas chapter of the NAACP, "350 paratroopers stood grimly at attention. Within minutes a world that had been holding its breath learned that the nine pupils, protected by the might of the U.S. military, had finally entered the 'never-never land.'"

"At 9:45 A.M. the Negroes crossed the threshold of school," was how one magazine began its description of events on September 25, 1957. But the conflict was not over. Above, one black Central High student fights back.

Warriors Don't Cry

Many of the quotations in this chapter are by then teenager Melba Pattillo. But long after she graduated from Central High, after she had a graduate degree from Columbia University and had worked as a reporter for NBC, Melba Pattillo (who was now Melba Pattillo Beals) wrote a book called Warriors Don't Cry. *It is a detailed telling of her extraordinary, and harrowing, experiences.*

I arrived at school Tuesday morning, fully expecting that I would be greeted by the 101st soldiers and escorted to the top of the stairs. Instead, we were left at the curb to fend for ourselves....

"Where are your pretty little soldier boys today?" someone cried out.

"You niggers ready to die just to be in this school?" asked another....

I wanted to turn and run away, but I thought about what Danny [a soldier in the 101st Airborne] had said: "Warriors survive." I tried to remember his stance, his attitude, and the courage of the 101st on the battlefield....Early that morning a boy began to taunt me as though he had been assigned that task.

First he greeted me in the hall outside my shorthand class and began pelting me with bottlecap openers, the kind with the sharp claw at the end. He was also a master at walking on my heels. He hurt me until I wanted to scream for help.

By lunchtime, I was nearly hysterical and ready to call it quits, until I thought of having to face Grandma when I arrived home. During the afternoon, when I went to the principal's office several times to report being sprayed with ink, kicked in the shin, and heel-walked until the backs of my feet bled...the clerk asked me why I was reporting petty stuff....

I thought a lot about how to appear as strong as I could as I walked the halls: how not to wince or frown when somebody hit me or kicked me in the shin. I practiced quieting fear as quickly as I could. When a passerby called me nigger, or lashed out at me using nasty words, I worked at not letting my heart feel sad because they didn't like me. I began to see that to allow their words to pierce my soul was to do exactly what they wanted.

Day after day, troops enforced the president's orders in Little Rock. "It is time," said one senator, "for the South to face up to the fact that it belongs to the Union and comply with the Constitution of the United States."

Pattillo. "I went in not through the side doors but up the front stairs, and there was a feeling of pride and hope that yes, this is the United States; yes, there is a reason I salute the flag; and it's going to be okay."

Ernest Green remembered the convoy that took him to school. There was a jeep in front and a jeep behind.

They both had machine gun mounts...the whole school was ringed with paratroopers and helicopters hovering around. We marched up the steps...with this circle of soldiers with bayonets drawn. ...Walking up the steps that day was probably one of the biggest feelings I've ever had.

At the end of that year, Ernest became the first black person to graduate from Central High. "I figured I was making a statement and helping black people's existence in Little Rock," he said. "I kept telling myself, I just can't trip with all those cameras watching me. But I knew that once I got as far as that principal and received that diploma, I had cracked the wall."

He was right.

"In some states—where people wanted to integrate but were afraid...they may go ahead now that they see they really have the backing of the federal government," said lawyer Thurgood Marshall (center).

17 Passing the Torch

On Inauguration Day, Jackie nearly stole the show—and her pillbox hat became a national fad.

Why was Kennedy, not Theodore Roosevelt, the youngest man ever elected to the presidency?

The United States is today the country that assumes the destiny of man....For the first time, a country has become the world's leader without achieving this through conquest, and it is strange to think that for thousands of years one single country has found power while seeking only justice.

—ANDRÉ MALRAUX, FRENCH POLITICIAN AND MAN OF LETTERS, ON A VISIT TO PRESIDENT KENNEDY, 1962

It had been cold all week in Washington, D.C., and Thursday night—January 19, 1961—snow fell thick and heavy. Washington, southern in its graciousness and geography, handles snow poorly. Everywhere cars stalled and people shivered.

That evening the army and navy were called. Three thousand servicemen, using 700 snowplows and trucks, worked through the night. The next day the wind was mean and the temperature stayed below freezing, but the streets were clear. Wooden bleachers were set up outdoors in front of the Capitol. At noon, when some 20,000 invited guests filled those bleachers, the winter sun reflecting off the banks of new snow seemed unusually bright. It was Inauguration Day.

Most of the presidential party wore scarves and mittens with their top hats and formal clothes. But the president-elect seemed to generate his own warmth—a quality he had in abundance. He took off his overcoat before he spoke. Then John Fitzgerald Kennedy put his hand on his grandfather's Bible and swore to uphold his mighty responsibilities. At 43, he was the youngest president since Theodore Roosevelt and the youngest man ever elected president. Next to him stood 70-year-old Dwight D. Eisenhower, at the time the oldest man ever to be president.

John F. Kennedy aged seven (right), with his sister Rosemary.

The contrast between the two was as strong as the winter sun. Genial, likable Ike was the son of a poor midwestern creamery worker. But his easy manner was an outside face; inside was a core of steel. Eisenhower had worked his way through an army career to the nation's top job.

Harvard-educated JFK, the patrician son of a wealthy businessman, had been given every advantage our society has to give. But the silver spoons that fed him had not made him lazy. Quite the opposite. He, and the other members of the large Kennedy family, were trained to serve their country, to achieve, and to do their best.

The wind blew as Robert Frost, America's favorite poet, read part of a poem he had written for this occasion:

> *Summoning artists to participate*
> *In the august occasions of the state*
> *Seems something for us all to*
>   *celebrate....*

A Chip off

P. J. Kennedy

John F. Fitzgerald and Patrick J. Kennedy (both sons of 19th-century Irish immigrants) became politicians. Fitzgerald, known as "Honey Fitz," was Boston's mayor. Kennedy, a saloon keeper (that means he owned a bar), was a senator in the Massachusetts legislature. But when he ran for the U.S. Senate, Henry Cabot Lodge, Sr. (an old-guard Bostonian) defeated him. (Lodge had also defeated Woodrow Wilson's dream of an American-supported League of Nations.) Years

Old Blocks

later, a grandson of Honey Fitz and P. J. Kennedy ran for the Senate against Henry Cabot Lodge, Jr.—and beat him. The new senator was John Fitzgerald Kennedy.

Honey Fitz

And then came lines about *a Golden Age of poetry and power, of which this noonday's the beginning hour.*

A "Golden Age of poetry and power." Would this handsome young president bring it about? There were many who believed he could. Not since the first days of Franklin Roosevelt's New Deal had so many eager people clamored to join the political process. The new cabinet (the president's top advisers) was going to be *bipartisan.* That meant it would include people from both political parties. Some of Kennedy's college professors were leaving their classrooms to become government officials. Thousands of Americans wanted to be part of the excitement that seemed to be building. It was amazing, the number of people who hoped to work to help their nation. John Kennedy had already suggested a "peace corps," a volunteer agency that would let Americans unselfishly share their experience and knowledge with less fortunate nations.

The young president, with his intense blue eyes, his

To *clamor* means to agitate noisily for something.

The huge extended Kennedy family almost always vacationed in a bunch at the family's summer house on Cape Cod. Here, Jack and nephews at the beach.

A Peace Corps worker in North Borneo. By 1963, about 5,000 volunteers were doing two-year stints in more than 40 Third World countries.

In 1961 the nuclear submarine *Triton* surfaced after making the first underwater around-the-world trip. *Triton* was named for the god of the sea in ancient Greek mythology. It had taken the sub 83½ days to complete its trip. (That was three days longer than Jules Verne's fictional hero Phineas Fogg took in what adventure book?)

thick head of hair, and his engaging smile, stepped up to the lectern and began to speak. "We observe today not a victory of party but a celebration of freedom," he said in strong, self-confident New England tones.

Let the word go forth from this time and place, to friend and foe alike, that the torch has been passed to a new generation of Americans, born in this century, tempered by war, disciplined by a hard and bitter peace, proud of our ancient heritage, and unwilling to witness or permit the slow undoing of those human rights to which this nation has always been committed....Let every nation know, whether it wishes us well or ill, that we shall pay any price, bear any burden, meet any hardship, support any friend, oppose any foe to assure the survival and the success of liberty....Let us begin anew...remembering on both sides that civility is not a sign of weakness.

Then he challenged his listeners:

If a free society cannot help the many who are poor, it cannot save the few who are rich....And so, my fellow Americans, ask not what your country can do for you, ask what you can do for your country.

After the applause, which was long and strong, the new president joined his wife, Jacqueline. "Oh,

A Solitary Child

Ecology—which comes from the Greek word meaning habitation— is the scientific study of our home: the earth. Here are some words about a famous ecologist named Rachel Carson.

Rachel Carson was, in her own words, "a solitary child." Brought up in Springdale, Pennsylvania, she spent "a great deal of time in woods and beside streams, learning the birds and the insects and flowers." When she was young, Rachel loved to read and thought she would become a writer. Then she decided to be a scientist, and at first believed that meant giving up writing. But of course it didn't have to mean that at all. She wrote of science and the natural world, and she did it so well that all who read her books gained a new awareness of their environment. Although, at first, no one paid much attention to what she wrote.

Then, in July of 1951, Oxford University Press (see the name on the spine of this book) published Rachel Carson's book *The Sea Around Us*. Oxford didn't expect much in the way of sales. What would you think if you were publishing a book about the ocean? There were hardly any humans in the book; it was all about reefs and islands and sea creatures and coral and sea plants. Would you think many people would read it? Oxford printed a modest number of copies.

The publisher was quickly astonished (and out of books). *The Sea Around Us* became a best seller—a huge best seller. The *New York Times* called it "the outstanding book of the year." Eventually it was translated into 32 languages. It introduced the ideas of ecology and conservation to large numbers of people. It was enormously influential.

By the end of the '60s, at least five state legislatures, alarmed by Rachel Carson's picture of a poisoned world, had banned or limited the use of DDT.

"We live in a scientific age; yet we assume that knowledge of science is the prerogative of only a small number of human beings, isolated and priestlike in their laboratories. This is not true. The materials of science are the materials of life itself. Science is part of the reality of living; it is the what, the how, and the why of everything in our experience," said Rachel Carson.

"It is impossible to understand man without understanding his environment and the forces that have molded him physically and mentally," she wrote. Then she attempted to explain that environment. Here is an excerpt from *The Sea Around Us:*

The Hawaiian islands, which have lost their native plants and animals faster than almost any other area in the world, are a classic example of the results of interfering with natural balances. Certain relations of animal to plant, and of plant to soil, had grown up through the centuries. When man came in and rudely disturbed this balance, he set off a whole series of chain reactions.

Vancouver brought cattle and goats to the Hawaiian Islands, and the resulting damage to forests and other vegetation was enormous. Many plant introductions were as bad. A plant known as the pamakani was brought in many years ago, according to report, by a Captain Makee for his beautiful gardens on the island of Maui. The pamakani, which has light, windborne seeds, quickly escaped from the captain's gardens, ruined the pasture lands on Maui, and proceeded to hop from island to island....

There was once a society in Hawaii for the special purpose of introducing exotic birds. Today when you go to the islands, you see, instead of the exquisite native birds that greeted Captain Cook, mynas from India, cardinals from the United States or Brazil, doves from Asia, weavers from Australia, skylarks from Europe, and titmice from Japan. Most of the original bird life has been wiped out.

The Sea Around Us made Rachel Carson famous; the last book she wrote, *Silent Spring*, brought her enemies (among some powerful interest groups). It took courage to write that book. It was a look at a grim subject—pesticides—and how they were poisoning the earth and its inhabitants. In *Silent Spring*, Carson attacked the chemical and food-processing industries, and the Department of Agriculture.

They lost no time in fighting back. Rachel Carson was mocked and ridiculed as a "hysterical woman." Her editor wrote, "Her opponents must have realized... that she was questioning not only the indiscriminate use of poisons but the basic irresponsibility of an industrialized, technological society toward the natural world."

But the fury and fervor of the attacks only brought her more readers. President Kennedy asked for a special report on pesticides from his Science Advisory Committee. The report confirmed what Carson had written, and it made important recommendations for curtailing and controlling the use of pesticides.

The public, which had been generally unaware of the danger of the poisons sprayed on plants, was now aware. Modestly, Rachel Carson said that one book couldn't change things, but on that she may have been wrong.

Above, John F. Kennedy, Sr. and Jr., at a Veterans Day ceremony. Below, the new president delivers his inaugural speech, hatless and coatless. His dislike of hats did serious damage to the hat industry.

Jack, what a day," she whispered. And it was. On that bright January afternoon, hope vibrated in the air. Our president was intelligent and graceful and knew how to laugh, especially at himself. He had big dreams that everyone could share. He intended that the nation reach for greatness within itself. He was surrounding himself with men and women who would be called "the best and the brightest." He expected to get things done.

And so, when he noticed in the inaugural parade that there were no black cadets among the Coast Guard marchers, his first act as president was to call an aide and ask him to do something about that. The next September there was a black professor and several black cadets at the Coast Guard Academy.

This man, John F. Kennedy, was determined to be an active president, a good president, a president who would inspire the nation. That wouldn't be difficult for him. Everyone talked of his *charisma* (kuh-RIZ-muh), which the dictionary says is *a rare power given to those persons with an exceptional ability for leadership and for securing the devotion of large numbers of people.*

18 Being President Isn't Easy

When Fidel Castro took over in Cuba, many Americans felt that democracy had a chance. But Castro soon embraced (literally) the Soviet leader Khrushchev.

The small island of Cuba, near Florida, had terrible problems. Its government was corrupt, criminals were making fortunes on the island, and most Cubans were very poor. So when Fidel Castro came along and took charge, in 1959, many Cubans and many Americans were hopeful. But when they learned that Castro was a communist—a Marxist-Leninist-Soviet communist—most stopped cheering. Castro did clean up much of the corruption in Cuba, improve the schools, and better race relations. But he was a dictator. The Cuban people were not free to oppose him or his ideas.

Many Cubans fled Cuba for the United States. They didn't intend to stay here. They wanted to go back to their own country and overthrow Castro. Most Americans would have liked that. The idea of a communist nation with ties to Russia sitting 90 miles off the coast of the United States made people in this country very nervous.

As soon as John F. Kennedy became president, he learned that the CIA—the Central Intelligence Agency—had been secretly training Cuban refugees as warriors and planning to land them on the island. The CIA experts said that the Cuban people would then rise up, join the invaders, and throw out dictator Castro. President Kennedy didn't want to seem soft on communism. After Joseph McCarthy, no one did. He told the CIA to go ahead.

The invasion, at a place called the Bay of Pigs, was a fiasco (fee-ASS-ko), which means it was a flop, fizzle, bomb, washout, dud, botch,

The CIA (Central Intelligence Agency) was established by Congress in 1947 and is America's foreign sleuthing operation. CIA agents are apt to be spies, or foreign intelligence gatherers. During the Cold War, the CIA became very influential, employing thousands of agents overseas and many others at its headquarters in Langley, Virginia (near Washington, D.C.). It spent its large budget without congressional scrutiny. (See what you can find out about what the CIA does today. Do you think we still need a spying organization?)

Can you find out why the Bay of Pigs is so named?

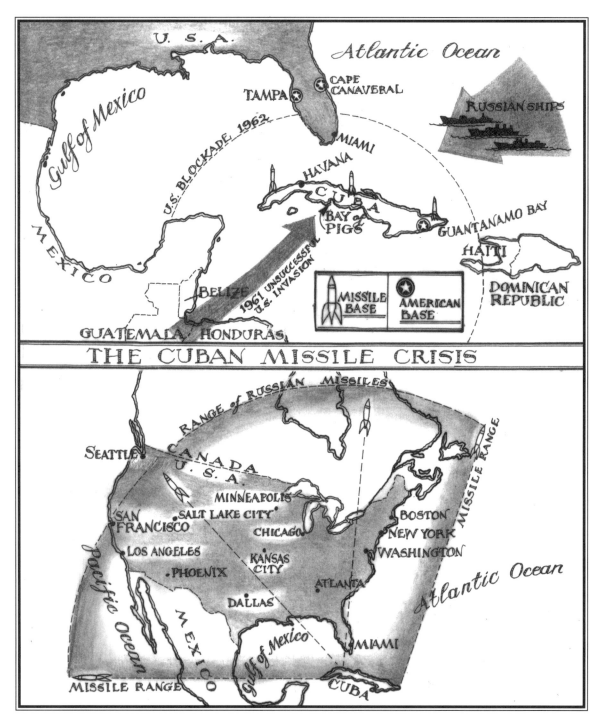

bungle, failure. Nothing worked right. The invaders were captured; the Cuban people didn't rise up; America—and Kennedy—looked foolish.

The young president took all the blame himself. "There is an old saying that victory has 100 fathers and defeat is an orphan," he said. "I am the responsible officer in this government," he added. The American people admired his honesty. His ratings in the popularity polls soared.

A U.S. U-2 spy plane took this aerial photograph of a medium-range nuclear missile site that was being built by the Soviets in Cuba.

But around the world people wondered, "Is he strong enough to be president?" Russia's leader, Nikita Khrushchev, was sure he wasn't. It seemed a good time to bully the United States. Premier Khrushchev decided to do something bold. He decided to put nuclear missiles in Cuba. They would be aimed right at America's most important cities and military targets. Some missiles were already in place when spy planes flying over Cuba brought news that missile sites were being prepared. Then Russian ships were sighted carrying more missiles.

What should the president do? The wrong move could start World War III. Both Russia and the United States had weapons that could destroy the world as we know it.

What would you do? The joint chiefs of staff (our top military leaders) wanted to bomb Cuba. Kennedy said no to his experts. He would not drop the first bomb, but he did announce that American troops were ready to invade the island if the missiles were not removed.

Khrushchev was in a tough spot, too. Castro wanted the Russians to launch a missile at the United States. Khrushchev said no to that. But he did tell the Russian military experts in Cuba that they could use nuclear weapons if there was an invasion.

Kennedy was firm. He said the missiles had to be removed. He gave Khrushchev time to make a decision. Secretly, Kennedy agreed to remove American missiles from Turkey. (Missiles there were a threat to Russia.) For 13 days the world held its breath, wondering if there would be a nuclear war. Then the missile-carrying Russian ships turned around and sailed home. The Cuban missiles were removed. The crisis was over.

But everyone knew that no one could win a nuclear war. Kennedy wanted both Russia and the United States to sign a treaty to stop testing nuclear bombs. We began talking with Russia about disarmament (reducing or doing away with weapons). Then, suddenly, Russia announced

On October 22, 1962, President Kennedy announced that he was blockading all ships carrying offensive weapons to Cuba. Would Russia back down?

The president and his closest adviser, his brother Bobby, during the missile crisis.

A **lobby** is a group that tries to persuade politicians to do what it wants. Some people think lobbies are a danger (most give congressmen money for campaign expenses). Others say that lobbying for your own interests is what democracy is all about.

A left-handed pitcher from Havana University tried out for the Washington Senators (now the Minnesota Twins), but the club turned him down. His name? Fidel Castro. (History is full of "what ifs." What if Castro had become a ballplayer?)

that bomb tests would begin again. Those tests put radioactive particles into the atmosphere—and that poisoned the air. Kennedy appealed to the United Nations. But Russia's tests continued. Finally, Kennedy said that the United States would have to begin testing again. Then he made a great speech at American University in Washington. The president said:

> Let us reexamine our attitude toward the Soviet Union....The wave of the future is not the conquest of the world by a single...creed but the liberation of the diverse energies of free nations and free men.

He was telling the Russian people that we needed to work together, not continue as enemies. Some Americans couldn't imagine getting along with Russia. Kennedy spoke to them:

> Some say it is useless to speak of world peace. I realize that the pursuit of peace is not as dramatic as pursuit of war...but we have no more urgent task....we all inhabit this small planet. We all breathe the same air. We all cherish our children's future. And we all are mortal.

A few weeks later, Khrushchev accepted the U.S. proposal for a test-ban treaty. "This treaty...is particularly for our children and grandchildren," said Kennedy, "and they have no lobby in Washington."

Meanwhile, things were still a mess in Vietnam. Remember, governments in the north and south were fighting for power. North Vietnam was backed by communist China and communist Russia; we were sending aid to South Vietnam. The North seemed to be winning. President Eisenhower had told Kennedy of the situation, "This is one of the problems I'm leaving you that I'm not happy about. We may have to fight."

President Kennedy wanted to know what was going on. He sent his vice president, Lyndon Johnson, on a fact-finding trip. When Johnson returned, he said the same thing that almost all our military and State Department experts were saying: if we wanted to see the communists defeated we would have to send more money and more supplies and more experts to train the South Vietnamese. So JFK sent the first American troops (called "advisers") to Vietnam. Except for a few lonely dissenters, no one asked if it was right to fight communists in Vietnam. Just about everyone in the early 1960s seemed to think we had to. By 1963 we had 11,000 military advisers in Vietnam; we were spending a million and a half dollars a day supporting that war.

Meanwhile, the struggle for civil rights continued in the American South.

On August 5, 1963, the Soviet premier, Nikita Khrushchev, toasted the signing of the Test Ban Treaty between the United States, the Soviet Union, and Britain.

19 Some Brave Children Meet a Roaring Bull

On Good Friday, 1963, Martin Luther King, Jr., led 50 hymn-singing marchers toward Birmingham's city hall, chanting "Freedom has come to Birmingham!"

It was hot, very hot, in the summer of 1962 in Birmingham, Alabama. But that didn't seem to make any difference to the city's white leaders. They closed all the city's public recreational facilities because they didn't want to see them integrated. That meant 68 parks, 38 playgrounds, six swimming pools, and four golf courses were locked up, and no one in Birmingham—kindhearted or mean-spirited, young or old—could enter the parks or swim in the city pools. For the wealthy, there were private pools and clubs, but for most people, there was no escaping the heat.

Bull Connor

Birmingham, Alabama's largest city, had plenty of moderate, clear-headed citizens, but the South's moderates were used to keeping quiet. Perhaps they feared mob action, or the disapproval of some of their friends, or the violence of the Ku Klux Klan. The Klan had helped elect Eugene "Bull" Connor as Birmingham's commissioner of public safety (police chief). Connor was about as big a bully as the South has ever produced. Besides that, he wasn't very smart. Bull Connor helped the civil rights movement a whole lot, although that wasn't what he intended to do.

In 1960, a *New York Times* reporter wrote about Birmingham, Alabama: "Whites and blacks still walk the same streets. But the streets, the water supply, and the sewer system are about the only public facilities they share."

95

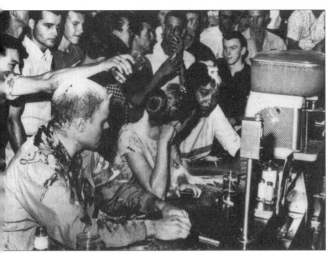

Students and a professor stage a sit-in at a segregated Woolworth's lunch counter in the state capital, Jackson, in 1963, while white youths·drench them with soda, ketchup, and mustard.

When Henry Louis Gates, Jr., applied for admission to Yale University, he wrote: "My grandfather was colored, my father was Negro, and I am black." What did he mean by that? Later, after he'd grown up and become a Harvard professor, Gates wrote in a letter to his daughters, "In your lifetimes, I suspect, you will go from being African Americans, to 'people of color,' to being, once again, 'colored people.'…I don't mind any of the names myself. But I have to confess that I like 'colored' best, maybe because when I hear the word, I hear it in my mother's voice and in the sepia tones of my childhood."

This was what happened: Birmingham's black citizens were marching, protesting, and demonstrating. They wanted the same rights as everyone else. They wanted to be able to eat in any restaurant. They wanted an end to segregation. They wanted to vote. Those were all their civil rights. They were demonstrating peacefully and nonviolently, but Bull Connor threw them in jail.

Martin Luther King, Jr., came to Birmingham, joined the marchers, and he was thrown in jail. Around the nation, people began to be concerned. A southern jail was a dangerous place for a black civil rights leader. King decided to write a letter from Birmingham jail and explain the reasons behind the civil rights movement. He didn't have any writing paper so he wrote on the margins of a newspaper and on toilet paper. He chose his words carefully. His powerful and persuasive letter explained what the marches were all about. He said it was unjust laws.

But Dr. King and the other leaders knew something dramatic was needed to capture the nation's attention. Demonstrators were being sent to jail every day, yet no one was doing anything about it. Thous-

Firemen turn high-pressure hoses on Birmingham civil rights demonstrators. "Every channel of communication," said a reporter, "has been fragmented by the emotional dynamite of racism, reinforced by the whip, the razor, the gun, the bomb, the torch, the club, the knife, the mob, the police."

"All you gotta do is tell them you're going to bring the dogs," Birmingham police chief Bull Connor told the press in May. "Look at 'em run. ...I want to see the dogs work."

ands of demonstrators were needed. Most blacks knew they would lose their jobs if they marched. Where could they find thousands of people who would march and not worry about losing their jobs?

In the schools.

"We started organizing the prom queens of the high schools, the basketball stars, the football stars," said Reverend James Bevel. Those student leaders got others interested. "The first response was among the young women, about 13 to 18. They're probably more responsive in terms of courage, confidence, and the ability to follow reasoning and logic. Nonviolence to them is logical....Then [came] the elementary students. The last to get involved were the high-school guys, because the brunt of violence in the South was directed toward the black male."

Bevel had already helped organize a demonstration in Nashville, Tennessee. College students there sat down at lunch counters and politely asked for items on the menu. When they weren't served, because of the color of their skin, they stayed in their seats—until the police took them off to jail. Across

The Highest Respect for Law

Martin Luther King, Jr., wrote a letter from the Birmingham jail. He addressed it to eight clergymen—Christian ministers and a Jewish rabbi—who had criticized the civil rights demonstrations and wondered why Dr. King had come to Birmingham. Here is part of what King said.

I am in Birmingham because injustice is here....I cannot sit idly by in Atlanta and not be concerned about what happens in Birmingham. Injustice anywhere is a threat to justice everywhere....What affects one directly affects all indirectly. There are two types of laws: just and unjust. I would be the first to advocate obeying just laws....

One who breaks an unjust law must do so openly, lovingly, and with a willingness to accept the penalty. I submit that an individual who breaks a law that conscience tells him is unjust, and who willingly accepts the penalty of imprisonment in order to arouse the conscience of the community over its injustice, is in reality expressing the highest respect for law.

the South, black and white students took part in *sit-ins* at lunch counters. Some people poured ketchup on the students' heads; others hit and kicked them. President Kennedy said, "The new way for Americans to stand up for their rights is to sit down." (But privately, JFK was urging Dr. King to ease up on the confrontations. He didn't want trouble.)

In Birmingham, boys and girls from the high schools, junior highs, and elementary schools wanted to march. "We held workshops to help them overcome the crippling fear of dogs and jails, and to help them start thinking on their feet," said Bevel, who taught the children the ways of nonviolence.

In Montgomery, when a woman was asked why she got involved in the bus boycott, she had said, "I'm doing it for my children and grandchildren." In Birmingham, Martin Luther King, Jr., said, "The children and grandchildren are doing it for themselves."

Some 600 children there marched out of church singing, and Bull Connor arrested them all. The next day another 1,000 children began a peaceful march. Connor called out his police dogs. Firemen turned on high-pressure hoses. The fire hoses were so strong they ripped bark off trees. When the water hit the children they were thrown on the ground and rolled screaming down the street. Television cameras hummed and people, worldwide, saw what was happening to Birmingham's children. Police dogs bit three teenagers so badly they had to be taken to the hospital. A small girl and her mother who knelt to pray on the steps of City Hall were arrested and taken to jail. Seventy-five children were squeezed into one cell built for eight prisoners. They sang freedom songs.

Patricia King was one of the children. She said:

> Some of the times that we marched, some people would be out there and they would throw rocks and cans and different things at us. I was afraid of getting hurt, but still I was willing to march to see justice done.

The Reverend Shuttlesworth was thrown against a wall by the powerful hoses; he had to be taken to a hospital in an ambulance. Bull Connor laughed when he heard that, and said he was sorry it wasn't a hearse.

Can you see why Connor helped the civil rights movement? Decent people were outraged. Most hadn't realized how bad things were for blacks in the segregated South. Now they could see for themselves on TV. In Washington, President Kennedy remembered stories his grandfather Honey Fitz had told him about anti-Catholic mobs who burned Catholic houses in 19th-century Boston. Kennedy knew that racial and religious hatred had no place in America. He asked his brother, Attorney General Robert "Bobby" Kennedy, to work to bring justice to all Americans. Bobby Kennedy would devote much of his energy to that cause.

On Sunday, September 15, 1963, a bomb exploded with the force of 12 sticks of dynamite during Sunday school at Birmingham's 16th Street Baptist Church. Above, relatives at the funeral of one of the four girls killed by the blast.

20 Standing With Lincoln

"We are not going to stop until the walls of segregation are crushed," said King. "We've gone too far to turn back now."

Former Brooklyn Dodger Jackie Robinson with his son at the March on Washington.

The civil rights leaders were human, and so there were rivalries and jealousies. They disagreed among themselves. Those from the older organizations, like the NAACP (National Association for the Advancement of Colored People), were at their best working through the courts and trying to change the laws. That was a slow process; it took skilled leadership. The lawyer Thurgood Marshall and labor chief A. Philip Randolph were that kind of leader.

NAACP

Martin Luther King, Jr., had helped organize the SCLC (the Southern Christian Leadership Conference). Its appeal was to the mass of moderate churchgoing blacks; most of its leaders were ministers. But many young people were impatient with both of these approaches, which seemed too slow-moving. They formed the Student Nonviolent Coordinating Committee (SNCC), known as SNICK. SNCC and the Congress for Racial Equality (CORE) organized many of the sit-ins in college communities.

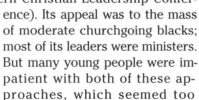

SCLC

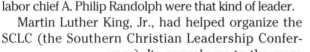

CORE

SNCC

Some black groups wanted to fight with fists, weapons, and anger. Everyone knew that if they got their way, much of the high purpose of the civil rights movement would be lost. Leaders like Martin

"You will never know how easy it was for me because of Jackie Robinson," said Martin Luther King, Jr., to Dodger pitcher Don Newcombe as the two of them ate dinner together one evening. "I never forgot those words," Newcombe remembered later. "It's a shame today when I ask a young ballplayer, or a young black kid I'm counseling on drug and alcohol abuse, who Jackie Robinson was and he can't tell me. It was Jackie and then it was me and Campy [Roy Campanella]....I make sure I talk about Jackie wherever I go. He was my idol, my mentor, my hero."

A group organized by CORE for the March on Washington gets ready to board the buses. "If I ever had any doubts before, they're gone now," said one marcher. "When I get back tomorrow I'm going to do whatever needs to be done."

"We Shall Overcome" became the anthem of the civil rights movement. The song is said to have originated in the 1940s at Tennessee's Highlander Folk School, where black textile workers gathered together.

We shall overcome,
We shall overcome,
We shall overcome
  someday.
Oh, deep in my heart,
  I do believe,
We shall overcome
  someday.

Luther King, Jr., had made civil rights a cause for all Americans. It was about equality. It was about justice and freedom for all. It wasn't just for blacks—although most of the leadership was black.

For years, A. Philip Randolph had talked of a freedom rally in the nation's capital. Perhaps it would bring the diverse black leaders together. Perhaps it would bring black and white people together. Perhaps it would influence Congress.

President Kennedy had sent a civil rights bill to Congress. Would it be passed? No one was sure. A march would show Congress and the president the importance of the civil rights movement. Many thought that Kennedy was paying more attention to affairs in Cuba and Vietnam than to the problem of unfairness at home. When President Kennedy gave a speech in West Berlin, Germany, about political freedom, it inspired cheers from people around the world. But some Americans weren't enthusiastic. They knew there was a kind of freedom that was missing right here in America—it went straight to the soul and spirit of an individual. The black leaders understood that soul freedom.

Exactly 100 years had passed since Abraham Lincoln signed the Emancipation Proclamation. Some white people were still telling black people to be patient. Martin Luther King, Jr., said, "We can't wait any longer. Now is the time."

Philip Randolph was 74. If ever he was to have his march, it had to be soon. And so it was decided; on August 28, 1963, there would be a march for freedom in Washington, D.C. Black leaders hoped that 100,000 people would participate. The marchers were going to demand four things: passage of the civil rights bill; integration of schools by year's end; an end to job discrimination; and a program of job training. Bayard Rustin, who

Bayard Rustin (left) and A. Philip Randolph

was a whiz at organizing, was put in charge.

Rustin got to work. He had 21 drinking fountains, 24 first-aid stations, and lots of portable toilets set up on Washington's grassy Mall. Workers made 80,000 cheese sandwiches. Movie stars, singers, high-school bands, preachers, and politicians practiced speeches and songs. The speakers and entertainers were to stand on the steps of the Lincoln Memorial and look toward the tall, slender Washington Monument and, beyond that, to the nation's Capitol.

Rustin worried about every detail. He got a big hook and put it on the end of a long stick. Then he gave careful instructions to a helper he called the *hook man*. Anyone who spoke too long was to be pulled from the microphones by the hook man.

Two thousand buses headed for Washington, and 21 chartered trains. A man with a freedom banner roller-skated from Chicago. An 82-year-old man bicycled from Ohio. Another, who was younger, came by bike from South Dakota. Sixty thousand whites came. Television crews, high in the Washington Monument, guessed that there were 250,000 people altogether.

It was a day filled with song, and hope, and good will. Finally, in the late afternoon, the last of the speakers stood on the steps of the Lincoln Memorial. It was Martin Luther King, Jr. He began with a prepared speech, which was formal and dignified, as was his nature. Then something happened inside him. Perhaps he responded to the crowd. Perhaps his training as a preacher took over. Whatever it was, he left his written speech and began talking from his heart. *"I have a dream,"* he said.

> *I have a dream that one day down in Alabama...little black boys and black girls will be able to join hands with little white boys and white girls as sisters and brothers.*
> *I have a dream today!*

Then he challenged the whole nation, not just those who were marching.

> *So let freedom ring from the prodigious hilltops of New Hampshire, let freedom ring from the mighty mountains of New York, let freedom ring from the heightening Alleghenies of Pennsylvania, let*

Less than a year after the March, groups of volunteer workers, black and white, most of them students, arrived by the busload in Mississippi on the Freedom Summer project. Their mission: to encourage blacks to register to vote. Three of them—James Chaney, a black Mississippian, New Yorkers Michael Schwerner and Andrew Goodman—disappeared, murdered by the Ku Klux Klan of Philadelphia, Miss. The killers were never convicted of murder.

"I have a dream," said Dr. Martin Luther King, Jr. "I have a dream that my four little children will one day live in a nation where they will not be judged by the color of their skin, but by the content of their character. I have a dream today!"

freedom ring from the snow-capped Rockies of Colorado; let freedom ring from the curvaceous slopes of California. But not only that. Let freedom ring from Stone Mountain of Georgia; let freedom ring from Lookout Mountain of Tennessee; let freedom ring from every hill and molehill of Mississippi. From every mountainside, let freedom ring.

And when this happens and when we allow freedom to ring, when we let it ring from every village and every hamlet, from every state and every city, we will be able to speed up that day when all God's children, black men and white men, Jews and gentiles, Protestants and Catholics, will be able to join hands and sing in the words of the old Negro spiritual: "Free at last. Free at last. Thank God Almighty, we are free at last."

What made King the most powerful and extraordinary black leader of this century was not his race but his morality. —SHELBY STEELE

21 The President's Number

"Kennedy could grasp an idea as fast as he could deal with figures," said an adviser. **"He absorbed ideas voraciously—and...permanently."**

The president didn't have much interest in arithmetic, said an aide, although he spent "most of his time counting." What President Kennedy was counting was votes in Congress, and the numbers never seemed to add up. Congress was controlled by an alliance of northern Republicans and southern Democrats. They voted as a bloc—against change. Kennedy, a liberal Democrat, wanted to take the country in new directions. The conservative Congress stood in his way. "When I was a congressman, I never realized how important Congress was. But now I do," moaned the president, with wry humor.

The nation had enjoyed quiet calm during the Eisenhower years. Now Kennedy had action in mind. Besides, after recessions in the later Eisenhower years, the economy needed a shove. And there was that continuing problem of unfairness. Most people thought that the American dream meant that everyone should have a chance to get a good education, to have decent housing, and a job. Some people weren't getting that chance. The president spoke to the American people of a "New Frontier" that would go beyond FDR's New Deal.

Kennedy had legislation that he wanted passed: civil rights bills, tax-cut bills, and health-care bills. There were also bills on equal pay for women, aid to the cities, aid to poor rural areas, manpower training, and a minimum wage. At first, the president was frustrated. Some people called his ideas "socialist." Fear of change, especially in the field of civil rights, caused his popularity to drop way down.

Zipping Around

What does ZIP stand for? Zone Improvement Program. ZIP codes weren't around when Benjamin Franklin was the first postmaster general. The five-digit numbers that tell the post office where your house is began under JFK's postmaster general, J. Edward Day. "The U.S. Post Office was handling more than half the world's mail in 1963," said Day. "Anybody could see that a machine could comprehend the meaning of 70631 easier than it could locate and direct a letter to Anaktuvuk Pass, Alaska." To get people to start using the codes, the singer Ethel Merman recorded a special version of the song "Zip-A-Dee Doo-Dah" from the movie *Pinocchio*. It seems to have worked.

A **recession** is a period of decline in economic activity. It is not as bad as a *depression*.

Rational means reasonable or sensible.

He seemed, in his own person, to embody all the hopes and aspirations of this new world that is struggling to emerge.

—HAROLD MACMILLAN, PRIME MINISTER OF BRITAIN IN 1963

But by 1963, the president saw signs that his ideas were being heard. John F. Kennedy seemed to know how to inspire people. He was optimistic, funny, intelligent, and forceful. Remember that charisma? It was working, and not just with Americans. Premier Khrushchev was blustering around and making loud statements, but later he wrote, "It quickly became clear that he [Kennedy] understood...that an improvement in relations [with Russia] was the only rational course."

In Congress, the numbers seemed finally to be adding up. The president was beginning to count enough votes to pass his bills. And Kennedy believed that the next election might bring him a Congress with even bigger favorable numbers. He expected to win a second term in the presidential elections in 1964. He hoped Congress would support his New Frontier. Except for that situation in Vietnam, things were looking good.

But there was trouble in Texas—political trouble—and Texas's votes would be important in the coming election. The Texas Democratic leaders couldn't seem to get along with each other. So when Vice President Lyndon B. Johnson, a Texan, asked Kennedy to go on a peacemaking mission, the president felt it his job to go.

Kennedy's press secretary got a letter. DON'T LET THE PRESIDENT COME TO DALLAS, it said. Texas is too dangerous, the letter writer added. The secretary put the letter aside. Everyone knew there was a group of noisy hatemongers in Dallas, but this president didn't seem to worry. And so, on a day filled with sunshine, he and Jackie waved goodbye to their two children and flew off to the Lone Star State.

Things started wonderfully well. The crowds in San Antonio and

Houston and Fort Worth were unusually warm and encouraging. At Fort Worth, unexpectedly, Kennedy went out into a parking lot and shook as many hands as he could. It bothered the Secret Service men, who were there to protect him, but the president liked contact with people.

Later in the day, when *Air Force One*, the presidential plane, touched down at Love Field in Dallas, thousands of people were waiting to cheer the president and the first lady. Mrs. Kennedy was handed a big bouquet of roses and asters.

The weather was so fine and the crowds so enthusiastic that the plastic bubble top was taken off the presidential car. The bulletproof side windows were rolled down. Texas's governor, John Connally, and his wife sat in the front seat of the big car; the Kennedys sat behind. They were on their way to a luncheon, and they took the busiest route through the city, so they could see the most people. Crowds lined the streets: a few among them were protesting, but most were cheering. When the limousine passed an old, seven-story schoolbook warehouse, Mrs. Connally

One witness recalled, "The movements in the president's car were not normal. Kennedy seemed to be falling to his left. ...There were two more explosions...only seconds after the first....People along the street were scattering in panic."

Outside Dallas's Parkland Hospital, Jackie, her stockings bloodstained, gets into the ambulance carrying her husband's body to *Air Force One*. "I don't think I ever saw anyone so much alone in my life," said Lady Bird Johnson, the vice president's wife.

Left, Mrs. Kennedy at the funeral. Right, John-John salutes his father's cortège. It was his third birthday. "I could actually hear people crying above the sound of the drums," said a member of the funeral drum corps. "There was nothing but the muffled drums and the tears."

turned around and said to the president, "You can't say that Dallas isn't friendly to you today!"

But President Kennedy never answered. Two bullets had pierced his head.

For the rest of their lives, most Americans would remember exactly where they were on November 22, 1963, when they heard the news. Again and again, they would stare at their TV screens and see the motorcade, the president falling into his wife's lap, the press at Parkland Hospital, and Jacqueline in her bloodstained suit.

At 1 P.M., John F. Kennedy was pronounced dead. At 2:30 P.M., Lyndon B. Johnson was sworn in as chief executive on *Air Force One*. That plane carried him, and the martyred president, back to the nation's capital. And the world wept.

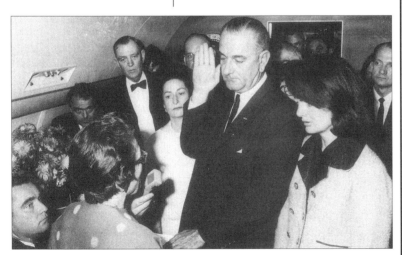

Lyndon Johnson is sworn in as 36th president of the United States. "He was extraordinary; he did everything he could to be magnanimous, to be kind," said Jackie. "I almost felt sorry for him, because I knew he felt sorry for me."

Ruby's Revenge

Millions were watching on TV as Jack Ruby (right) shot Lee Harvey Oswald (center) on his way to the county jail.

Dallas police capture Lee Harvey Oswald in a movie theater after he shoots a policeman who notices his suspicious behavior. Oswald, who has lived in Russia and Cuba, is charged with the shooting of President Kennedy and Governor Connally (who was wounded). Two days later, going through a jail passageway, Oswald is shot and killed by Jack Ruby, a Dallas nightclub owner. The debate about whether others were involved in JFK's assassination continues to this day.

22 LBJ

LBJ's Texas ranch, near Johnson City, where he grew up, became the "Western White House"; the president would pretend to rope cattle for the cameras.

The new president was big. Taller than six feet three inches, he had big bones, big ears, a big nose, big hands, and big feet. His voice was big, his ego was big, and when it came to his ambition—it was bigger than big. His ambition was colossal.

He wanted to be a great president, right up there with Washington and Lincoln. No, he wanted more than that. He said he wanted to be "the greatest of them all, the whole bunch of them."

Johnson's vice president, Hubert Humphrey, said later:

> *He was an all-American president. He was really the history of this country with all of the turmoil, the bombast, the sentiments, the passions. It was all there, all in one man, and if you liked politics, it was like being at the feet of a giant.*

Johnson's dream was to wipe out poverty in America. He wanted to see blacks, whites, Hispanics—all people—treated as equal citizens. He wanted old people to be cared for. He wanted no barriers to hinder the handicapped. He wanted every child in the country to get a good education. He wanted to see an America where *all men [and women] are created equal*. And he worked for those goals with more energy and political savvy than any president before or since. He understood, as few have, that helping the poor and the disadvantaged would enrich the whole nation.

Lyndon Baines Johnson came from Texas, from the scruffy Hill Country near Austin, a region so isolated when he was a boy that no

Ego (EE-go) is a person's sense of self.

LBJ aged about 5

Turmoil is chaos.
Bombast is big, loud, bragging talk.
Sentiments are thoughts.

Those LBJ initials were a family thing. The first lady was Lady Bird Johnson and the president's daughters were Lynda Bird and Lucy Baines.

Lyndon's father, Sam Ealy Johnson, in the state legislature in Austin. Father and son's best times together were spent campaigning for Sam's reelection. "We drove in the Model T Ford from farm to farm, up and down the valley, stopping at every door," LBJ recalled.

one had electricity at home, and almost no one had running water indoors, or an indoor toilet. If you wanted to take a bath, or do the dishes, or wash clothes, you had to pump water from a well in the yard, carry it inside the house, and heat it over a fire. At the first school Lyndon went to, all the grades were in one room, with just one teacher. Most of the children didn't wear shoes. But the Hill Country people didn't think of themselves as poor. They had food to eat and roofs over their heads—and they knew and cared about each other.

Lyndon was bright—everyone could see that—but he was a rebellious student. Sometimes he did well; often he didn't. A tutor who was called to help him remembered that he was "Mischievous! That boy was terrible." A cousin said, "Lyndon would always have to be the leader."

From the time he was a little boy it was politics that fascinated him. That wasn't surprising. His mother was the daughter of a Texas secretary of state. His father, Sam Ealy Johnson, served in the Texas legislature. Everyone knew

Lyndon Johnson (middle row, center) with the Mexican-American children of Cotulla, Texas. As teacher he taught the kids to play ball games; he made them speak English in school, and recite and debate and compete in spelling bees. He was very popular with both children and parents. "It was like a blessing from a clear sky," said one pupil later on.

Sam Ealy. He wore a six-shooter on his hip, a cowboy hat on his head, and he tucked his pants into high boots. He was tall, loud, boastful, and sometimes mean. He could also be very loving. His son both feared and adored him.

By the time Lyndon was six he was attending political rallies and handing out pamphlets. When he was ten he would go with his father to the legislature and "sit in the gallery for hours watching all the activity on the floor." He knew what he wanted to do with his life. "I want to wind up just like my daddy, gettin' pensions for old people," he told a friend.

But when he was ready to go to college, his father was in debt. The family farm had failed. Lyndon borrowed $75 to help pay his expenses. After a year he had to drop out and teach in order to earn money to finish. He taught Mexican-American children that year and saw real poverty—worse than anything he knew. He never forgot those children.

Back at college, he got a job carrying trash and sweeping floors. "He made speeches to the walls he wiped down, he told tales of the ancients to the doormats he was shaking the dust out of," a friend said. Others remembered that he was "always in high gallop," or "clamoring for recognition." It was his energy everyone talked about. He never seemed to stop. If he was asked to do a job, he did double the job. "It pained him to loaf," said another friend. He ended up working as an assistant to the college president, who later laughed and told him, "You hadn't been in my office a month before I could hardly tell who was president of the school—you or me."

When Lyndon Johnson graduated, the college president said, "I predict for him great things in the years ahead." But he had no idea he was talking to a future president of the United States.

When young LBJ first arrived in Washington (right, as a congressional assistant, outside the Capitol), said an observer, he was "tall as a plow horse and slim as a lodgepole; he boasted a slick of black curly hair and a smile as wide as the Pedernales [River]—a very attractive guy so cocksure of himself that he never stopped talking."

When we say, "One Nation under God, with liberty and justice for all," we are talking about all people. We either ought to believe it or quit saying it.
—HUBERT H. HUMPHREY

A ***pension*** is retirement pay or old-age insurance.

23 The Biggest Vote in History

FDR and LBJ meet in Galveston, Texas. In the center is Texas governor James Allred; later on he was often airbrushed out of this picture to enhance Johnson's importance.

Some people are born to be preachers and some to be teachers and some to be ballplayers. Lyndon Johnson was born to be a politician. He was 29 when he was first elected to Congress, and he set out like a sprinter in a running race. Right away, he arranged to meet President Franklin Roosevelt. Afterward, FDR called an assistant and said, "I've just met a most remarkable young man...this boy could well be the first southern president [in a century]."

As a congressman, Johnson worked 16-hour days and sometimes longer; he expected his aides to work right along with him. That was nothing new for him, but one of his assistants had a nervous breakdown. Naturally, Johnson got a lot done, and impressed some Washington old-timers. One of them recalled, "There was this first-term congressman who was so on his toes and so active and overwhelming that he was up and down our corridors all the time." Another remembered, "This fellow was a great operator....Besides the drive and the en-

Andrew Johnson, who was president right after the Civil War, and who also succeeded a martyred leader, was the last southern president until Lyndon Baines Johnson.

[Johnson] said the only power he had was the power to persuade. That's like saying the only wind we have is a hurricane.

—RALPH HUITT,
A SENATE COMMITTEE ASSISTANT

Lyndon and Lady Bird. Their Alabaman friend Virginia Durr (see page 78) said: "Bird was a sweet-looking, dark-haired, dark-eyed girl who seemed to adore her husband....She talked very little and let him do all the talking."

ergy and the doing favors, which he did for everybody, there was a great deal of charm in this man." All that energy meant benefits for his district. He got the government to help finance slum-clearance projects and low-cost housing in Austin, the state capital. And he insisted that Mexican-Americans and blacks have a fair share of the new houses. Money he got for the region helped farmers go from back-breaking horse-and-plow farming to 20th-century farm machinery. He brought electricity to the Texas Hill Country.

Johnson (left) at Austin's Santa Rita Housing Project. He fought for federal money to build these apartments, and he won; Austin was one of the first five cities in the country to get FHA (Federal Housing Authority) approval for a new kind of loan issued through the Public Works Administration.

Farming wives soon had washing machines. Farming families could turn on the lights. "Of all the things I have ever done," said Johnson 20 years later, "nothing has ever given me as much satisfaction as bringing power to the Hill Country of Texas."

A congressman represents one district in a state. Some states have many congressmen; the number depends on the state's population. But each state has only two senators; each senator represents the whole state. Twelve years after entering Congress, Lyndon Johnson was elected to the Senate. Four years after that, he was elected leader of the Democratic Party in the Senate. A cyclone in the Senate chamber might have been less noticed. When Lyndon Johnson let loose with his never-stop Texas energy, he usually got whatever it was he wanted. But he wasn't all bluster. He knew how to compromise. During the Eisenhower presidency he worked closely with the Republicans. He helped them get bills passed; they helped him get favors for Texas.

President Kennedy's programs had been stalled in the Republican Congress. President Johnson knew how to trade and maneuver and twist arms. He soon began to get Kennedy's programs passed, and then he added his own vision—it extended the New Frontier and the New Deal. He called it the *Great Society*. It was a vision of a place where there was no poverty; where all children were well schooled; where health care was a birthright; where jobs and job training were attainable by all. If he could make all that happen—well, he would be the greatest president.

The American people didn't know quite what to make of Lyndon Johnson. Sometimes he just seemed like a big, crude country boy who had stumbled into the presidency. Some people made fun of him. The

The Great Society is not a safe harbor, a resting place, a final objective, a finished work. It is a challenge constantly renewed, beckoning us toward a destiny where the meaning of our lives matches the marvelous products of our labor.

—LYNDON JOHNSON, 1965

111

Kennedys had served gourmet meals to Harvard professors, artists, and Nobel prize-winners in the White House dining room. The Johnsons seemed to be serving Texas chili to politicians.

Johnson was hurt. He wanted everyone to admire and love him. He didn't want to be an accidental president. He wanted to be elected president himself. He soon had that chance. In 1964 he took his ideas to the American people. He ran for the presidency on his own. He didn't just want to be elected; he wanted the

Johnson worked for years to get the Mansfield Dam approved and built. It controlled floods on the Lower Colorado River and supplied hydroelectric power too.

The Hill Country, where Johnson came from, was one of the poorest, most isolated parts of Texas. It had no effective electricity as late as 1937. That didn't just mean no lights or refrigerators; it meant no running water, because you needed electricity to pump the water. Johnson got the government to electrify the Hill Country; he changed its farmers' lives.

biggest popular vote in the history of the country.

Well, he got what he wanted: the biggest popular vote ever! And he also got a Congress that was Democratic; it would support his programs. Now he had an opportunity few presidents have had. He had that grand vision of a Great Society. He had support from the people to get things done. And he had the ability and energy to make it happen.

The American people had elected him so overwhelmingly that he could be himself. He decided he would wear a gray business suit to his inauguration instead of the traditional top hat, tails, and striped pants. He even danced at his inaugural ball. According to the record books, only two presidents had done that: George Washington and William Henry Harrison. Isaac Stern (a great American violinist) and Van Cliburn (a great American pianist who happened to be from Texas) performed with the National Symphony Orchestra at an inaugural party. At a State Department reception, composers, writers, and dancers were honored. The president was triumphant.

But he had already done something that would tear down much of what he had carefully constructed. He had told the American people and Congress that the North Vietnamese had attacked an innocent American ship in the Gulf of Tonkin. That wasn't quite true, and he knew it. (If you keep reading, you'll hear more about the Gulf of Tonkin—and that untruth.)

January 8, 1964.
President Johnson tells Congress that he is declaring a war on poverty. He outlines a plan that includes aid to Appalachia, youth employment programs, improved unemployment insurance, a domestic Peace Corps, and expansion of the area redevelopment program. Later, his budget puts the cost of the war on poverty at $1 billion. Actual expenditures in fiscal 1964–1965 are slightly more than $600 million. The deficit for that year is the lowest in five years. (A government has a *deficit* when it spends more money than it brings in— from taxes, import duties, etc. What is a *fiscal year*?)

This administration, here and now, declares unconditional war on poverty in America....It will not be a short or easy struggle, no single weapon or strategy will suffice, but we shall not rest until that war is won.
—LYNDON JOHNSON, 1964

"All the way with LBJ" was Johnson's election slogan. He shook so many hands on his 1964 campaign that he had to be bandaged to protect his own bruised, bleeding hand.

24 Salt and Pepper the Kids

Often, a president's hardest job is making Congress pass the laws and programs he wants. This cartoon portrayed Johnson as a brilliant dealmaker who could get Congress to play any tune he desired.

As new discoveries are made, new truths disclosed...institutions must advance also, and keep pace with the times.

—THOMAS JEFFERSON

Hi-fi is short for *high fidelity*—meaning true sound reproduction. It was what everybody called stereo systems when they first became popular in the late '50s and early '60s.

In 1721, an English writer named Jonathan Swift wrote a satire. (Swift is the author of *Gulliver's Travels,* and if you haven't read that book, try it. You will like it.) *Satire* is literature that uses wit, humor, and irony to expose wickedness. Benjamin Franklin loved satire, and wrote it too.

In case you are wondering why an 18th-century English satirist is in a book about 20th-century America, here is the explanation.

England was prosperous in the 18th century. Most Englishmen and women thought they were living in good times. But they couldn't understand why there were so many poor people in England and Ireland. They decided that the poor must be to blame for their own problems.

The same kind of thing happened in 20th-century America. The United States was thriving. We were called an "affluent society." That means we were rich. Many Americans had cars, bikes, television sets, hi-fi sets, and nice houses. But some people were left out. Some people went hungry. They didn't get a fair chance to go to a good school or to get a good job. Were they to blame for that?

What should be done about poverty? That problem was as old as society, and not easy to solve. In 18th-century England and Ireland, many poor families and children struggled and starved. Some, who were convicts, were just shipped off to America. Jonathan Swift suggested something else in his satire, which was called *A Modest Proposal for Preventing the Children of Ireland From Being a Burden to Their Parents or Country.*

What did Swift propose to do to keep poor children from being burdensome? His solution was simple: eat them!

> *A Child will make two Dishes at an Entertainment for Friends, and when the Family dines alone, the fore or hind Quarter will make a reasonable Dish and seasoned with a little Pepper or Salt will be very good Boiled on the fourth Day, especially in Winter.*

Does that sound grisly? Remember, this was satire. Swift wanted to make people think. He wanted them to see how silly some of their ideas were. Jonathan Swift thought something real could be done about poverty, and he wanted to shock people into thinking about it.

Lyndon Johnson also thought something could be done about poverty, and he meant to do it. He intended to build his Great Society. It would take money, the talent to get bills passed in Congress, and the leadership to make Americans understand that ending poverty would make everyone richer.

Johnson revved up his jet-engine personality and blasted away. He

It is part of the American character to consider nothing as desperate; to surmount every difficulty by resolution and contrivance.
—THOMAS JEFFERSON

The 24th Amendment, outlawing the poll tax as a device to prevent citizens from voting, is ratified by South Dakota on January 23, 1964, completing approval by three-quarters of the states.

According to the Census Bureau, between 1960 and 1990, poverty was reduced by 39 percent—from 22.2 percent of the nation's population to 13.5 percent. That was still too many poor people, but many fewer than there would have been without anti-poverty programs.

Johnson tours poor areas of Appalachia during his election campaign. (His own roots were in a poverty-stricken part of Texas.) "Our aim is not only to relieve the symptoms of poverty," he said, "but to cure it, and above all, to prevent it."

In 1964, the 24th Amendment outlawed poll taxes as a requirement for voting in federal elections. Some feared—as this cartoon shows—that the amendment would make it harder to get the Voting Rights Act passed, but in 1965 it was done.

Do you know which of Lyndon Johnson's agencies and programs still exist? Do some of the ones described here help you or people you know?

Rosemary Bray grew up in Chicago on welfare. But her mother made her children study and she helped them do well in school. Rosemary got a scholarship to Yale University, graduated in 1976, and became a writer. She says that her mother, and the anti-poverty programs started by LBJ, saved her from a life of penury, which is exactly what they were supposed to do.

got a *civil rights act* passed—it outlawed most discrimination. *Operation Headstart* helped little children prepare for kindergarten. The *Job Corps* found work for school dropouts. *Upward Bound* helped needy children go to college. The *Neighborhood Youth Corps* trained unemployed teenagers. The *Teacher Corps* trained schoolteachers. *Medicare* helped old people pay their hospital bills. *Medicaid* helped those without much money afford a doctor. Soon there was less poverty in America. There were very few homeless people.

Johnson went to the Statue of Liberty to sign a bill that ended narrow, racist immigration quotas. Because of an immigration bill passed early in the century, only people from some privileged regions, like western Europe, were able to come to the United States in large numbers. Just a few from other regions—Asia, for instance—were allowed to enter and become citizens. Johnson's new law let new groups of immigrants (especially Asians and Latinos) broaden the American family.

The president began beautification programs and environmental-protection programs. Congress and the president worked together. FDR had accomplished a lot during his first 100 days in the presidency; Johnson got even more laws passed.

All the new programs cost money. But we had the money. We were a rich nation. Johnson knew that. We could afford the Great Society. We could afford the war on poverty—until something else began taking most of our money. That was the war in southeast Asia.

American soldiers were now fighting in Vietnam. But things weren't going the way we hoped they would go. North Vietnam was fighting back. That little nation didn't seem to be frightened by America's power.

Being president had turned out to be harder than Lyndon Johnson expected it to be. His

Johnson had a great sense of drama. He signed his immigration bill beneath the Statue of Liberty; he signed an education bill in the one-room schoolhouse he had attended; he signed the Voting Rights Act in the Rotunda of the Capitol in Washington, D.C.

"I do not find it easy to send our finest young men into battle," LBJ said about Vietnam. But he sent them anyway. Meanwhile, more and more people were turning against the war (left, at a demonstration in Des Moines), and the war's huge cost was making it hard to pay for Johnson's domestic reforms, as the cartoon illustrates.

military and political advisers kept telling him to send more men and more weapons to Vietnam. But some of America's citizens were saying that the war was a mistake. They thought we needed to quit the war. Johnson didn't like that advice at all; he didn't want to be a quitter.

And then there was the black community. President Johnson thought that African-American people, and other minorities, would be thrilled with the Great Society programs. They were pleased. But they weren't satisfied; they wanted more. They wanted what everyone else had. They wanted to be equal partners in America. Why, they even wanted to vote!

When it came to voting, Lyndon Johnson asked the black leaders to be patient. He explained that it wouldn't be easy to get a voting act through Congress. Change would have to come one step at a time, he said. But some people were fed up with being patient.

Hey, hey, LBJ,
how many kids
did you kill today?
—VIETNAM DEMONSTRATORS' CHANT

By 1967, the Vietnam War was costing the country $70 million a day.

LBJ's presidency began with the support of civil rights leaders such as Martin Luther King, Jr. (right), who said, "We're on our way to a Great Society." But the war eventually drove them apart.

25

A King Gets a Prize and Goes to Jail

Prize Giver

Alfred Nobel was born in 1833 to a Swedish family that manufactured torpedoes and mines and explosives. Nobel was trained as a chemist; he became an inventor. In 1866 Nobel perfected a new explosive; he named it *dynamite*. It brought him great riches. But he was disturbed that his wealth came from weapons of destruction.

In his will, he left a fund to provide for annual international awards—Nobel prizes —in physics, chemistry, physiology, medicine, literature, and for work promoting international peace. (In 1969 an award in economics was added.) The awards (some think them the most prestigious in the world) are decided by the Swedish Royal Academy of Science. They include a gold medal and a sum of money. Here are the names of a few Nobel winners. Can you find out who they are and what their awards were for? Rabindranath Tagore, Rudyard Kipling, Albert Einstein, Max Planck, Thomas Mann, Sinclair Lewis, Luigi Pirandello, Eugene O'Neill, Harold Urey, Pearl Buck, William Faulkner, Par Lagerkvist, Boris Pasternak, Ernest Hemingway, Juan Ramón Jimenez, Albert Schweitzer, Linus Pauling, Isaac Bashevis Singer, the Dalai Lama, Paul Samuelson, Andrei Sakharov, James Watson, Francis Crick, Joseph Brodsky, Mother Teresa, S. A. Waksman, Elie Wiesel, Gabriel Garcia Marquéz, Steven Weinberg, Nelson Mandela, Saul Bellow, Derek Walcott, Mikhail Gorbachev, Toni Morrison.

"Nonviolence is the answer to the crucial political and moral questions of our time," said Martin Luther King, Jr., in his Nobel acceptance speech.

Martin Luther King, Jr., was in the hospital. He wasn't seriously ill; it was a case of exhaustion. It was Tuesday, and he'd given three speeches on Sunday and two on Monday, and then there were all those trips to jail, and the marches, and the pressures. But when the phone rang he felt a whole lot better. Matter of fact, he felt great.

His wife, Coretta, had big news: Martin had been awarded the Nobel Peace Prize. That prize is given each year to the person, from anywhere in the world, who has contributed most to peace. Along with the great honor, there is a sizable cash award. Theodore Roosevelt had won the Nobel Peace Prize, and so had Woodrow Wilson and Jane Addams. Martin Luther King, Jr., at 35, was the youngest person ever to receive it.

Some Americans were furious, and they wrote to the Nobel committee in Sweden and told them so. Bull Connor said, "They're scraping the bottom of the barrel." Some racists called it a communist plot. But most Americans were proud. Newspaper columnist Ralph McGill, writing in the *Atlanta Constitution*, said that Europeans understood King better than most Americans; they saw in him "the American promise," with its message for the whole world.

King flew to Europe to receive the Peace Prize. He invited his parents and his wife and Bayard Rustin and Ralph Abernathy—25 friends in all—to go with him. In England, he gave a sermon at London's most famous church, 291-

year-old St. Paul's Cathedral. He was the first non-Anglican to preach there. The chairman of the Norwegian parliament said he was "the first person in the Western world to have shown us that a struggle can be waged without violence." (Gandhi was part of the Eastern world.) When Norwegian students sang, "We Shall Overcome," King realized that his message had become part of a universal language of freedom.

King was soon back in the United States—and in jail again. He was in Selma, Alabama, trying to help black citizens vote.

President Johnson's Civil Rights Act didn't solve the voting problem. It did allow black people to check into any hotel they desired, sit on buses wherever they wished, and eat in any restaurant. But in much of the rural South, blacks still couldn't vote. In 1964, when blacks tried to register to vote in Alabama or Mississippi or some other southern states, they were likely to be beaten, or to lose their jobs—even though the 15th Amendment to the Constitution says that every citizen has the right to vote. Those who kept trying to register were given impossible questions to answer, or asked to pay a poll tax they couldn't afford.

Selma was the seat of Dallas County, Alabama. The county offices were located in its courthouse. It was a town of 30,000 people; more than half were black. Selma was Bull Connor's birthplace. It was an Old South cotton town on the banks of the Alabama River. Before the Civil War one of the town's buildings had been used to hold slaves—sometimes 500 of them—as they waited to be auctioned off. During the Civil War, Selma was a Confederate military depot. In the 1960s, the streets in the black section of town were made of red dirt; those in the white section were paved. Mule-drawn carts still clattered down Selma's dusty streets, hauling cotton.

The leaders of Selma's black community asked Dr. King (and his organization, the SCLC) to come to town.

Registering to vote under official supervision. In 1964 only two-fifths of the South's eligible black population was registered.

A demonstrator campaigns for the vote and equal employment outside the Florida state capitol in Tallahassee.

We are a great nation, I think, largely because of our protection of the right to criticize, to dissent, to oppose, and to join with others in mass opposition—and to do these things powerfully and effectively.

—ABE FORTAS, ASSOCIATE JUSTICE, U.S. SUPREME COURT

Martin Luther King gave his Nobel prize money to the civil rights movement.

I have sworn upon the altar of God eternal hostility against every form of tyranny over the mind of man.

—THOMAS JEFFERSON

Marchers on their way to Montgomery, Alabama.

In the Selma jail, King discovered a man who had been locked up for more than two years—and still hadn't been told why he'd been arrested. That was against the Constitution. It was horrible. It was not unusual in some areas of the South.

In 1965, when King and the SCLC arrived, SNCC workers had already been in Selma for more than a year. They had worked hard to try to get blacks signed on the voting rolls. SNCC had doubled the number of registered black voters to 333 (out of 15,000 of voting age).

A few SNCC members weren't happy that King was asked to come to Selma. They wanted to stay in charge. But John Lewis, SNCC chairman, was not one of them. He was thrilled that Martin Luther King, Jr., was coming to town. So were the majority of Selma's citizens.

Dallas County Sheriff Jim Clark wasn't thrilled—he was angry. Clark was a big, crude, blustery white guy who wore a military jacket, carried a nightstick, hated Negroes, and said so. Clark was an embarrassment to Selma's leading white families, but he'd been elected sheriff anyway. (If the county's black citizens had been voters it wouldn't have happened.)

Martin Luther King, Jr., spoke out at Brown's Chapel (an old, sturdy, red-brick church with two steeples and an outside balcony). "Give us the ballot," he cried. A group of Selma's black citizens marched to the courthouse to try to register. They weren't allowed inside. Sheriff Clark made them stand in an alley. When SNCC workers tried to bring them sandwiches and water, the workers were hit with billy clubs. That just made Selma's black citizens more determined.

More than 100 black teachers marched. Teachers usually steer clear of controversy, but this time they didn't. They wanted to vote. The teachers' march was the real turning point, said the Reverend Frederick Reese. "The undertakers got a group, and they marched. The beauticians got a group; they marched. Everybody marched after the teachers marched," said Reese.

Martin Luther King, Jr., marched with 250 citizens who wanted to register to vote. They were all thrown in jail. King, too. When they heard of Dr. King's arrest, 500 schoolchildren marched to the courthouse. They were arrested. Two days later 300 more schoolchildren were arrested. The evening television news covered it all. King wrote a letter from jail. He said, "This is Selma, Alabama. There are more Negroes in jail with me than there are on the voting rolls." Fifteen congressmen came to Selma. They announced that "new legislation is going to be necessary." President Johnson held a press conference and said, "All Americans should be indignant when one American is denied the right to vote."

Coretta Scott King went to the jail to visit her husband. She brought a message from Malcolm X, who was in Selma. Malcolm, a black leader who was electrifying urban audiences with hard facts and a spirit of militancy, had been invited to Selma by SNCC's leaders. His ideas were different from King's. Malcolm had never recognized the power and force of nonviolent action. But Malcolm seemed to be heading in a new direction. He told Coretta, "I want Dr. King to know that I didn't come to Selma to make his job difficult." Then he added, "If the white people realize what the alternative is, perhaps they will be more willing to hear Dr. King."

The alternative was violence. Speaking to a big crowd in Brown's Chapel, Malcolm said, "White people should thank Dr. King for holding people in check, for there are others who do not believe in these [nonviolent] measures."

Malcolm had broken with the Black Muslims who believed in the separation of blacks and whites. He'd gone to Mecca in Saudi Arabia—the center of the Muslim world—where he converted to orthodox Islam. That experience moved him deeply. It caused him to change some of his ideas—especially ideas that called for hate and violence. He told a journalist about that change in himself.

> *The sickness and madness of those [early] days. I'm glad to be free of them. It is a time for martyrs now. And if I'm to be one, it will be in the cause of brotherhood. That's the only thing that can save this country. I've learned it the hard way—but I've learned it.*

Two and a half weeks after his trip to Selma, Malcolm X was martyred—killed by Black Muslims—a victim of the violence that he had once tolerated. Malcolm's new belief in brotherhood made the loss of this brilliant Muslim especially tragic.

Malcolm X (above) urged black people to be proud of their blackness and their African roots. He had been a leader of the Nation of Islam, or Black Muslims, but broke with the movement over the advocacy of violence by some members. Below, women at a Nation of Islam meeting that proclaimed them property to be protected from contamination by whites.

Muslims are followers of Islam and its prophet, Mohammed. Islam, like Judaism and Christianity, is based on belief in one God. Mecca, in Saudi Arabia, is Mohammed's birthplace. It is considered Islam's holiest city and its religious center.

A ***martyr*** is someone sacrificed to a cause.

26 From Selma to Montgomery

The Montgomery marchers, led by Hosea Williams (left) and SNCC chairman John Lewis (right), try to cross the Edmund Pettus Bridge in Selma on "Bloody Sunday," March 7, 1965.

The tension in Selma was awful. Marchers, and even reporters covering the marches, were being roughed up and beaten. Where could they go for protection? Not to the police. The police, the state troopers, and Sheriff Clark were doing most of the beating. When 82-year-old Cager Lee marched, a state trooper went for him and whipped him until he was bloody. Jimmy Lee Jackson, Cager's grandson, carried his grandfather into a café.

But the troopers weren't finished; they stormed right into the café. One trooper hit Jimmy's mother, another shot Jimmy Lee Jackson in the stomach. He died seven days later.

That murder did it. It was too much for the civil rights workers to bear. They felt responsible. There was no stopping them now. "We had decided that we were going to get killed or we were going to be free," said one leader. The murder was also too much for some of Selma's white citizens. Seventy of them marched in sympathy to the courthouse. One white minister said:

> We consider it a shocking injustice that there are still counties in Alabama where there are no Negroes registered to vote....We are horrified at the brutal way in which the police at times have attempted to break up peaceful assemblies and demonstrations by American citizens.

What is a firebrand? See Book 3 of **A History of US** *for the story of three famous firebrands.*

Six hundred people—men, women, and children—gathered at Brown's Chapel. They were prepared to march the 58 miles from Selma to Alabama's capital, Montgomery. They intended to face Governor George Wallace and demand that all of Alabama's citizens

be protected in their right to vote.

Most of the 600 were black, although some were white. Hosea Williams, a young firebrand, was in charge. (Martin Luther King was in Washington consulting with the president.) First they prayed, and then they began their march, singing as they went. They marched six blocks from Brown's Chapel to the Edmund Pettus Bridge. No one stopped them; all was quiet except for their voices. They knew that once they crossed the bridge they would be on the road to Montgomery.

When they mounted the sloping crest of the bridge they were stunned by what they saw: Alabama state troopers were lined up, gas masks in place, bullwhips and billy clubs raised. The troopers didn't give anyone time to decide what to do. They moved forward; some were on horseback, some on foot. Then they released tear-gas bombs. Eight-year-old Sheyann Webb said:

> I saw people being beaten and I tried to run home as fast as I could....I saw horses behind me....Hosea Williams picked me up and I told him to put me down, he wasn't running fast enough.

But something new had come to this out-of-the-way southern town. That something was television coverage. Camera crews were filming the action. Sheriff Clark's bullying was not just Selma's problem; it was national news. Television stations across the nation interrupted their regular programs to show scenes of policemen on horseback clubbing peaceful marchers. "It looked like war," said Selma's mayor. "The

Leaders of the march: from far left, Bayard Rustin, A. Philip Randolph, John Lewis, Ralph Abernathy; third and fourth from right, Coretta Scott King and Martin Luther King, Jr.

Truth is great and will prevail if left to herself.
—THOMAS JEFFERSON

Ain't going to let no posse turn me 'round, Keep on walkin', keep on talkin', Marching up to Freedom Land.
—HYMN SUNG BY SELMA-TO-MONTGOMERY MARCHERS

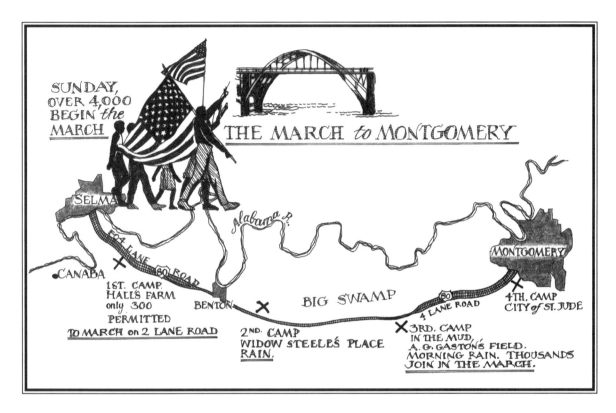

THE MARCH to MONTGOMERY

SUNDAY, OVER 4,000 BEGIN *the* MARCH

SELMA

•CANABA

1ST. CAMP. HALL'S FARM only 300 PERMITTED TO MARCH on 2 LANE ROAD

Alabama R.

BENTON

2ND. CAMP WIDOW STEELE'S PLACE RAIN.

BIG SWAMP

3RD. CAMP IN THE MUD, A. G. GASTON'S FIELD. MORNING RAIN. THOUSANDS JOIN IN THE MARCH.

4 LANE ROAD

MONTGOMERY

4TH. CAMP CITY *of* ST. JUDE

Rob a people of their sense of history and you take away hope.

—Rev. Wyatt T. Walker, aide to Dr. Martin Luther King, Jr.

Where are Lexington and Concord? Why are they famous? What about Appomattox?

wrath of the nation came down on us."

How would you feel if you watched all that on television? Most good people were sickened. So, when Martin Luther King, Jr., sent telegrams to prominent clergymen saying "Join me for a ministers' march to Montgomery," ministers came from many places and many faiths. White-bearded Rabbi Abraham Heschel came from the Jewish Theological Seminary, world leader Ralph Bunche came from the United Nations (he, too, had won a Nobel Peace Prize), and Unitarian minister James Reeb came from Boston.

Reeb did not go back to Boston. The nightmare of brutality wasn't quite finished in Selma. Reeb and some white ministers made the mistake of eating in a black café. For Reeb it was a fatal mistake. He was clubbed to death when he came out of the restaurant.

President Johnson was shocked, and said:

What happened in Selma was an American tragedy. At times, history and fate meet in a single place to shape a turning point in man's unending search for freedom. So it was at Lexington and Concord. So it was a century ago at Appomattox. So it was last week in Selma, Alabama.

The president announced that he was sending a voting rights bill to Congress. Then he spoke to the 70 million people who listened on television. *It's not just Negroes,* he said. *It's really all of us who must overcome the crippling legacy of bigotry and injustice. And,* he finished with these words from the civil rights theme song, WE SHALL OVERCOME.

"We were all sitting together," said a black leader who heard the president speak. "And Martin was very quietly sitting in the chair, and a tear ran down his cheek. It was a victory like none other."

Six days later, 4,000 people—black and white—marched from the Pettus Bridge in Selma to Montgomery, camping out at night and singing songs of freedom. This time National Guardsmen protected them. By the time they reached the capital, 25,000 people had joined the march. Rosa Parks was there, and so were many of those who, 10 years earlier, had walked through winter's bluster and summer's heat rather than ride Montgomery's segregated buses. Martin Luther King, Jr., had been an unknown preacher then. Now he was world famous.

Top, state troopers in gas masks (for protection against tear gas) meet marchers on the bridge; below, seconds later, troopers rush forward, knocking marchers down.

The 54-mile march took five days. "We are on the move now," said Dr. King. "Let us therefore continue our triumph and march...on ballot boxes until the Wallaces of our nation tremble away in silence."

27 War in Southeast Asia

By 1971—when this cartoonist drew "The Blind Leading the Blind"—four presidents had found themselves entangled in Vietnam. Can you name them all?

The war in Vietnam was costing billions of dollars a year. Someone said we had to make a choice between *guns* and *butter.* (Guns symbolized war and butter stood for goods and helping programs.) At first, President Johnson thought we could have both. But the guns got more and more expensive. Soon we were spending more in Vietnam than on all the welfare programs combined. Funds for the Great Society had to be cut. Many of its programs were eliminated.

The Vietnam War turned out to be a terrible mistake. We blundered into it without knowing much about southeast Asia. We never took time to learn.

It was a civil war. We made it our war. It became a battle between the most powerful nation in the world and a small country of farmers. It was bombers, helicopters, and rockets in a nation with water buffalo and barefoot runners. What were we fighting for? It was supposed to be for freedom and democracy. But since we hadn't done our homework and didn't know much about the country we were fighting in, we backed corrupt leaders in South Vietnam who robbed the treasury and bossed everyone around.

It is easy to see mistakes after you've made them; that is called *hindsight.* No nation wants to make mistakes. We didn't enter the Vietnam War in order to do wrong. It was that issue of communism that caught us. The North Vietnamese were getting money and supplies from communist China and from the Soviets, too. Many Americans feared that the Chinese communists would control a united Vietnam. Because we hadn't studied much Vietnamese history, we didn't know that the Chinese and the Vietnamese didn't get along very well.

Most of the advisers to Presidents Eisenhower, Kennedy, and Johnson believed we should fight in Vietnam. They believed it was

Henry David Thoreau had protested during the Mexican War. He had even gone to jail rather than pay taxes that would support the war. A few people protested with him. But not enough to make a difference. Besides, we won that war, and quickly too, which made it popular. Vietnam was different. It went on and on and on. And we didn't win.

America's role to stand up to any communist nation, anywhere. They believed that all communist nations were part of a large conspiracy. If Vietnam was allowed to become communist, everyone seemed sure that all of southeast Asia would soon follow.

Actually, the Vietnamese could have used our help. They were faced with a poor choice: between a miserable dictator and a repressive communist government. We might

The U.S. Army destroyed Vietnamese villages on purpose, to turn peasants into refugees. The refugees, like this woman carrying all her worldly goods, had to leave their homes and flee to the cities; the Americans believed this would make it harder for Vietcong soldiers to find food and support in the countryside.

have given them some guidance. Uncle Sam could have sent teachers and business advisers and our democratic ideas. But many of our 20th-century leaders talked about freedom and democracy, yet sometimes acted as if they didn't really believe in them. We didn't back free elections in Vietnam. Instead, we spent vast sums of money on weapons and we tried to solve problems the unthinking way—by fighting.

We got into the war in Vietnam one step at a time. It wasn't the fault of the Republicans or the Democrats. It was bipartisan. In the early 1960s, war hawks in both parties were screaming that we needed to fight. Our presidents didn't want to be called soft on communism. Besides, they remembered Hitler, the terrible tyrant who started World War II. If Hitler had been stopped early, that war might not have happened. But the leader of North Vietnam, Ho Chi Minh, was no Hitler (though he wasn't a democratic leader, either).

Our leaders didn't ask the right questions, and so we sent more than half a million Americans to do battle in a faraway land.

We got deeper and deeper into the Vietnamese jungle—and then we didn't seem to know how to get out. First there was that little step of Truman's. Then Eisenhower invested more money and sent more advisers. And Kennedy sent much more money and lots of advisers.

President Johnson didn't know what to do. His advisers were pushing him to enter the war in a big way. Barry Goldwater, who was the Republican candidate for president in 1964, ran a get-into-the-war campaign and even talked about using nuclear weapons. Johnson ran as the peace candidate. You remember Lyndon Johnson's big ego, though. He didn't want to look like a coward. Then something happened soon after he was elected that gave him an excuse to become a warrior.

Above, on a Saigon street, a scene later watched on TV by millions of Americans: the execution (by South Vietnam's police chief) of a man suspected of being a Vietcong guerrilla. Below, a guard threatens a captured Vietcong.

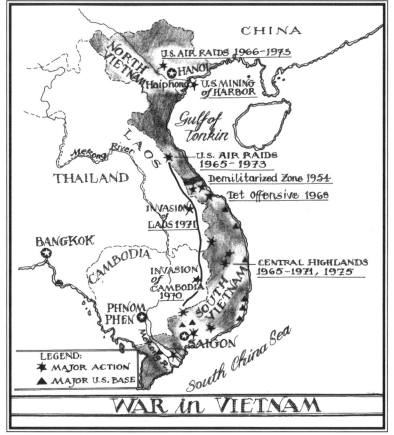

Map labels:
CHINA
NORTH VIETNAM
U.S. AIR RAIDS 1966-1973
HANOI
Haiphong — U.S. MINING of HARBOR
Gulf of Tonkin
LAOS
Mekong River
U.S. AIR RAIDS 1965-1973
Demilitarized Zone 1954
Tet Offensive 1968
THAILAND
INVASION of LAOS 1971
BANGKOK
CAMBODIA
INVASION of CAMBODIA 1970
CENTRAL HIGHLANDS 1965-1971, 1975
SOUTH VIETNAM
PHNOM PHEN
SAIGON
South China Sea
LEGEND:
★ MAJOR ACTION
▲ MAJOR U.S. BASE

WAR in VIETNAM

"Only Thing We're Sure Of—There Is a Tonkin Gulf!"

An American ship was on a secret mission in the Gulf of Tonkin. It wasn't supposed to be there. A torpedo was fired at the ship. Two days later there was a second report of torpedoes that turned out to be a mistake. Maybe a sailor saw a flying fish. President Johnson exaggerated things. He said an American ship had been attacked. He got Congress to pass a resolution that let him go to war. It was called the Gulf of Tonkin resolution.

We began bombing North Vietnam. Before we finished we dropped more bombs on that small country than we had on both Germany and Japan during all of World War II. And that wasn't the worst of it. There is something about war that no one wants to talk about. It is what soldiers with killing weapons sometimes do. Most of our soldiers were decent and many were heroic. Many helped the people of Vietnam. But some didn't. Imagine trying to fight in a hot jungle where you can't even see the enemy. Some soldiers were angry and violent, and they had terrible weapons and they used them on innocent villagers. The Vietnam War became a national nightmare. It went on and on and on. We destroyed much of the land, and we killed and were killed.

The Pentagon (which is the name for our military control center) just couldn't understand how guerrilla (say it like GORILLA) fighters who had their ammunition carried over jungle trails on the backs of old men and women could beat a modern army supplied by

Arkansas senator J. William Fulbright—pointing at map, left—opposed the war policies of LBJ and his advisers, such as Robert McNamara (pointing, right). Once friendly colleagues, Fulbright and Johnson stopped speaking to each other.

An Unwilling Guest at the Hanoi Hilton

It was July 18, 1965, and U.S. Navy Commander Jeremiah A. Denton, Jr., was sitting in the cockpit of a bomb-loaded A6 airplane on the carrier *Independence*, located in the Gulf of Tonkin. The deck of an aircraft carrier seems huge, but its runway isn't long enough for a normal takeoff. So the plane was catapulted—like an arrow from a bow—off the deck.

Denton was on his way to North Vietnam. As he dropped his bombs over the target, near Hanoi, he felt a jolt. His plane had been hit. The controls were dead. "My heart pounded. It was over. I was very frightened," he wrote later. Jeremiah Denton descended into the Ma River, was picked up by Vietnamese soldiers, and became a POW: a prisoner of war.

At home in Virginia, Denton's wife, Jane, and his seven children learned that he had been captured (other pilots had seen him eject from the plane). They didn't know that he was put in solitary confinement—which means he was all alone in a tiny cell with only a concrete bed, wooden stocks that held his legs, and a bucket for a toilet. They didn't know other things, either. That he was physically tortured, mentally tortured, starved, and taken on a march through Hanoi where people hit him and spat on him. Or that he sometimes lost all sense of who he was. Imagine if it were you. Could you handle it? For Denton, 1965 became 1966, and '67, and '68, and on, and on.

He was one of the first POWS, but eventually some 700 Americans were captured and held in several prisons near Hanoi. They gave the prisons names like the Hanoi Hilton, Dogpatch, Heartbreak, Alcatraz, and Briarpath. The prisoners kept their sanity and their pride in themselves and their country by defying their captors.

Jeremiah Denton as a hostage of war, in a Vietnamese film designed to show that hostages were well treated by their captors.

They maintained military discipline. As a high-ranking officer, Denton was often in command. How could he do that?

The POWS talked to each other by tapping on the walls of their cells with a special code. When the jailers stopped the tapping, they whistled or coughed the code. The tap code was based on this chart:

	1	2	3	4	5
1	A	B	C	D	E
2	F	G	H	I	J (K)
3	L	M	N	O	P
4	Q	R	S	T	U
5	V	W	X	Y	Z

Try it. To tap *A* you go 1-1. *B* is 1-2, *R* is 4-2, and *X* is 5-3. The letter *K* is 2-6.

The American people disagreed about the war—its purpose and necessity—but everyone was in agreement when it came to the POWS. We wanted them home. The Vietnamese soon realized that they had important hostages. They filmed some of the POWS, intending to show that they were being well treated. Commander Denton kept blinking his eyes. Most people thought it was the bright camera lights, but Denton was blinking the word *torture* in Morse code.

International law sets standards for the treatment of prisoners. Torture is forbidden. Four people who saw the film recognized the code and called government officials.

The North Vietnamese were not the only ones who tortured prisoners. The South Vietnamese army also committed terrible atrocities.

What happened to the POWS? How long were they guests at the Hanoi Hilton? Read on and you'll find out.

Children from Trang Ban, South Vietnam, flee after their school was burned with napalm during an American raid.

helicopters. The military chiefs kept telling the president that if we just sent a few thousand more soldiers and dropped a few more bombs it would all be over. But the old men and women and the guerrilla fighters, who seemed to know how to vanish into the jungle, finally made the great and mighty United States give up and go home.

We thought we were doing the right thing when we began. We really were unselfish. We weren't imperialists. We didn't want to make Vietnam a colony. And we left much of our national wealth in that nation halfway around the world. So why did we make such a terrible mistake?

We didn't understand what the war was all about.

It was about freedom. The Vietnamese wanted to be free of foreign rule. They wanted to choose their own leaders. They wanted freedom even to make the wrong decisions. This was a nasty civil war. We soon made it much worse. We made it a high-tech war. We brought in grenades, rocket launchers, jellied-gasoline explosives (called *napalm*), and chemicals (called *defoliants*) that took all the leaves off the jungle trees—and we still couldn't beat the Vietnamese.

We should have known that could happen. After all, we ourselves started out as a little pipsqueak nation that defeated the great and mighty British empire. Didn't we remember that people fighting for their own freedom are apt to be unbeatable? What had happened to us?

Dorothy Day (left) protests the war at a draft card–burning ceremony.

Going to Jail to Fight War

Dorothy Day was in her sixties when she stood outside a missile base and tried to block its entrance. She was protesting against an industry that built weapons of destruction. At the same time, she urged Americans not to pay their income taxes, because tax money supported the war in Vietnam. (Henry David Thoreau had done the same thing at the time of the Mexican War.) Day was willing to go to jail for her beliefs—and she did.

She believed that war was wrong. She thought that capitalism led people to concentrate on getting and spending money, and she didn't think that brought happiness. She thought that nationalism—love of country—often led to war and misery. So what did she believe in? The power of God's love. She believed that God could be found among the poor, and that helping the poor was a way to change society and find meaning in life.

Abbie Hoffman, a fiery young 1960s radical, called her "the first hippie," and Dorothy Day was proud of the label. She was the founder of the *Catholic Worker*, a newspaper devoted to educating the public about the plight of the poor and uneducated. The *Catholic Worker* sponsored soup kitchens and hospitality houses for the poor.

But how do you treat someone who breaks rules? Dorothy Day was a problem to some Catholic church members. They wanted to take official action against her; others found her inspiring. *Time* magazine didn't see a problem. It called her "a living saint," and put her on its cover.

130

28 Lyndon in Trouble

By 1967, Vietnam was a raw scar on the president's body. "I feel like a hitchhiker caught in a hailstorm on a Texas highway," said LBJ. "I can't run. I can't hide. And I can't make it stop."

Lyndon Johnson was miserable. He knew he was losing his dream of a Great Society. But he didn't know how to stop the war in Vietnam. He didn't seem able to admit that he had made a mistake. He had started with that fib about an attack in the Gulf of Tonkin. But you know how those things go: one lie usually leads to another, and sometimes another, and another.

That's what happened to President Johnson. He said, "We are not going to send American boys nine or ten thousand miles away from home to do what Asian boys ought to be doing for themselves," when he was already planning to do just that. He said all the bombing was "aimed at military targets." But newspaper reporters told of houses, schools, and stores flattened by bombs. President Johnson kept saying that we were winning the war and

it would soon be over. TV made people realize he wasn't telling the truth. For the first time in history, ordinary people could see exactly what war was like. The TV screen showed dead American soldiers and dead Vietnamese.

At first, it was mostly students on college campuses who began demonstrating against the war. Then more and more American people began to join them. Martin Luther King, Jr., was now leading anti-war protests, as well as civil rights marches. Ministers of many faiths were doing the same thing. The college protests began to get ugly and violent.

Nineteen sixty-seven was the height of the hippie era—the "summer of love." That year, even all-American *Life* magazine published an editorial that said the Vietnam War was no longer worth winning—and was too much "to ask young Americans to die for." Outside the Pentagon, some young Americans protested the war by planting flowers in the military policemen's guns.

In 1966, race riots in Atlanta (top) escalated into looting of stores; a year later, violence in Detroit (above, left and right) lasted a week, resulting in 43 deaths and 2,000 wounded.

And then the cities, especially those in the North, started exploding. America's cities had been neglected. In many, schools were terrible, transportation was terrible, crime made life frightening, and there weren't enough jobs for those who wanted to work. City people were fed up. That was a bad break for the president. He was trying to do something to improve the cities. But not everyone was willing to wait for his programs to work. Things had been too bad for too long. In 1965, riots in the Watts section of Los Angeles lasted six days and left 34 dead. Newark, Chicago, Cleveland, and other cities erupted with riots of their own. America's cities were like volcanoes filled with frustration and pain. The riots may have eased the frustration; they didn't do much to help the pain. Johnson asked Illinois governor Otto Kerner to head a commission to investigate the riots. The governor said they could be traced to "white racism." The Kerner Commission warned, "Our nation is moving toward two societies, one black, one white—separate and unequal."

Black protests began changing direction. New leaders appeared; many were angry young people. They had no patience with nonviolence. The new leaders didn't talk about brotherhood and love; they talked of power, separation, and sometimes hate.

Left, militant Black Power leader Stokely Carmichael speaks to college students. "We been saying freedom for six years," he said, "and we ain't got nothing." Inset, playwright LeRoi Jones, who led a voter registration campaign that put blacks in control of Newark, New Jersey.

132

Martin Luther King said of one militant group, "In advocating violence it is imitating the worst, the most brutal, and the most uncivilized value of American life." Anti-Vietnam protests grew louder and more strident. Many who had supported the war were now changing their minds.

Black Power voices were followed by Brown Power, or Mexican-American, voices. These were people who wanted their full rights as citizens. So did the female half of the population. Women had been demanding equal pay for equal work and not getting it. Women's rights leaders joined the protest fray. Some were just angry; others had clear ideas and programs.

Would you like to have been president in the '60s? By 1968, the country seemed to be coming apart.

Conscience of the Court

Thurgood Marshall (left) calls his wife from the Oval Office with the news that the president has appointed him to the Supreme Court.

Lyndon Johnson and Thurgood Marshall were friends. They were both energetic men, and talkers, and they understood each other. Each of them wanted to make the world better than it was. But there was one thing Lyndon Johnson told Thurgood Marshall he wasn't going to do. He wasn't going to make Marshall a Supreme Court justice. If Thurgood Marshall were a justice, he wouldn't be free to talk politics with the president. They couldn't be friends anymore.

So, in 1967, Marshall didn't know what Lyndon Johnson had in mind when he was invited into the Oval Office of the White House. The president was sitting at his desk. They chatted a bit, and then, according to Marshall, President Johnson said, "You know something, Thurgood?"

"No, sir," said Marshall. "What's that?"

"I'm going to put you on the Supreme Court."

And Marshall said, "Oh, yipe!" and "What did you say?"

But he had heard right. President Johnson had changed his mind. Thurgood Marshall was to be the first black justice ever appointed to the Supreme Court. After a while, after they talked some more, Marshall called his wife and told her to sit down because the president had something to tell her. After that, Johnson said, "I guess this is the end of our friendship."

"Yep," said the justice-to-be. "Be no more of that." Then the president recalled when a justice that Harry Truman had appointed made a decision that Truman opposed.

"You wouldn't do like that to me?" he said to Marshall, who answered: "No sooner than," which meant he certainly would do like that if he thought it the right thing to do. To which Johnson replied, "Well, that's the way I want it." And that is the way it was. Thurgood Marshall was an independent-minded justice, a great justice, a justice one lawyer described as the "conscience of the court."

By the end of 1967, only 26 percent of Americans approved of Johnson's handling of the Vietnam War.

A Farming Village in Vietnam

On the map, My Lai (me-LY) was a pinprick of a place on the northeast coast of South Vietnam, near the South China Sea. Early on the morning of March 16, 1968, when the men of Charlie Company of the 11th Brigade of the Americal Division entered the village, My Lai entered the annals of history. The American soldiers shot, at point-blank range, everyone they could find: old men, pregnant women, children, babies—347 civilians in all.

Lieutenant Calley

Then the soldiers ate lunch, went back, shot all the domestic animals—water buffaloes, pigs, chickens—threw the carcasses into the village wells (to poison the water), and burned the place down. Their officers didn't seem to think they had done anything wrong.

But a few of the soldiers had refused to participate, and Hugh Thompson, Jr., who was in a helicopter overhead, threatened to shoot the GIs if he saw them "kill one more woman or child." He landed between the GIs and the villagers.

A few miles away, soldiers in Bravo Company (of the same army division) wiped out a hamlet called My Khe in a similar bloodbath. Later, officers and men lied about what they had done. Only after a newspaper reporter (Seymour Hersh) wrote the truth (and won a Pulitzer prize)

did Americans at home understand what their soldiers had done.

Lieutenant William Calley, Jr., who had directed much of the massacre (and who personally killed 109 Vietnamese, including babies), was the only soldier convicted of a crime. He was court-martialed and sentenced to life in prison at hard labor. But President Nixon intervened (to his and America's shame), and Calley was released after three years of house arrest in his own apartment.

If Americans had any illusions about the war, My Lai shattered them. Writer Neil Sheehan said:

Had they killed just as many over a larger area in a longer period of time and killed impersonally with bombs, shells, rockets, white phosphorus, and napalm, they would have been following the normal pattern of American military conduct. The soldier and junior officer observed the lack of regard his superiors had for the Vietnamese....The military leaders of the United States, and the civilian leaders who permitted the generals to wage war as they did, made the massacre inevitable.

A long time ago, war was a gentleman's pursuit. It was often heroic. There were things in war to boast of, and tell your grandchildren about. But even then—in the days of knights and valor—there had always been the other side, which was brutal, gruesome, and disgusting. After Vietnam, and especially after My Lai, there was almost nothing left in war but the shame of it.

Quang Ngai Province, South Vietnam, March 16, 1968: the massacre at My Lai.

29 Friedan, Schlafly, and Friends

"For the first time in their history," wrote Betty Friedan, "women are becoming aware of an identity crisis in their lives, a crisis which began generations ago."

HELMER: *Before all else, you are a wife and mother.*

NORA: *That I no longer believe. I believe that before else, I am a human being, just as much as you are—or at least that I should try to become one.*

—HENRIK IBSEN, *A DOLL'S HOUSE* (1879)

Television programs (and newspaper and magazine articles) can tell you a lot about a time—or at least the way people of that time saw themselves. And in the '50s, the number-one TV program was *I Love Lucy*. Lucy was a dippy dame—a white, middle-class wife and mother—who didn't work, was kind of bored, and was always getting into mischief with her neighbor Ethel. One day, Lucy and Ethel decided to get jobs on the production line at a candy factory. As the chocolates passed by—zip-zip fast—Lucy and Ethel got farther and farther behind. So they began stuffing chocolates in their mouths and then in their dresses—and it was all a big laugh.

There were mixed messages in that *Lucy* show. Did women belong in the candy factory, or at home? Lucy and Ethel always seemed to be trying to break away from the household routine. But they usually goofed up. And behind the show was the real Lucille

Below, a typical day at home with Lucy and Ethel. In January 1953, more people saw *I Love Lucy* on Monday nights than watched President Eisenhower's inauguration in the same month.

"The suburban housewife," wrote Betty Friedan, "was the dream image of young American women and the envy, it was said, of women all over the world." Top, the '50s ideal: Mom and daughter look on admiringly as their lord and master mans the barbecue. Bottom, a young mother deals with reality. Right, two of TV's blissfully happy families: *Ozzie and Harriet* (top) and *Father Knows Best* (bottom).

Ball, who was not only one of the most gifted comedians this country has produced but a powerful businesswoman as well.

The reality of the '50s was that most middle-class white women—like it or not—did stay home. A lot of them lived in the new suburbs, in new houses, with new washing machines and a kind of life that seemed idyllic. Most TV programs told you that as long as Mom didn't venture out into the big world, her life was close to perfect. The shows were all full of happy suburban couples and cute kids: *Ozzie and Harriet, Leave It to Beaver*, and *Father Knows Best* (note that title) were three very popular shows that portrayed terrific, happy households.

Yet when writer Betty Friedan decided to do an article about suburban women, she found that many of them were not happy—and they didn't know why. She called it the "problem that has no name." Then she investigated, and said that women weren't being given a chance to develop their talents. They were taught to "keep their place." If a woman had the potential to be a brain surgeon, if she wanted to be a veterinarian, if she thought being an architect would fill her life with pleasure—well, too bad. Those, and most professions, were for males only. Women had a separate—and unequal—role in society. They were expected to stay in the kitchen and the nursery, perhaps get involved with community volunteer work, or maybe just putter around, look pretty, and be frustrated.

Friedan knew something about frustration. She had lost a reporting job when she got pregnant. And she couldn't get that article on discontented women published—magazines only seemed to want to tell about domestic bliss. So Betty Friedan decided to turn her research into a book. She called it *The Feminine Mystique*.

That book, published in 1963, put in words what a lot of women had been thinking. More than a million copies were sold. It made people think—some men as well as women. It carried an idea—that all people, including women, have the natural right to develop their potential. It told how the media (advertising, TV, radio, newspapers, and magazines) were manipulating women in order to keep them at home where they could be sold vacuum cleaners and dishwashers. Friedan's book came along at the same time that other things were happening to make women reconsider their role in society.

Suburban women may not have been working, but many other women were at work. It had begun during World War II (from 1941 to 1945), when about 6.5 million women who had not worked before got jobs. It was patriotic. Men were fighting; women were needed on production lines. Many women found they liked working. Some had no choice: they had to work.

After the war, many of those women kept working. Others began to work too. "By 1960," writes William H. Chafe, an expert in the field, "both the husband and the wife worked in over 10 million homes (an increase of 333 percent over 1940)." But attitudes didn't change. Working women were paid less than men. And they hardly ever got prestige jobs. Many were teachers, but women were not superintendents of schools. Many were nurses, but very few were doctors.

And those who worked had to put up with prejudice against working women. An article in the *Atlantic Monthly* was typical of the times. It said, "What modern woman has to recapture is the wisdom that just being a woman is her central task and her greatest honor."

But what the magazines and TV shows were saying was in conflict with what more and more women were beginning to believe: that they had minds and talents equal to men's.

Then, when the civil rights movement erupted in the '60s, women became some of its hardest workers. And they often found that the men in the movement expected them to make coffee, do the cleaning up, and not make major decisions. That was infuriating. The civil rights movement was all about equal rights. So some women took the activism they learned as civil rights workers and brought it to the women's liberation movement.

Many of those women, and others who called themselves radical feminists, didn't think Betty Friedan went far enough. Friedan wanted women to be equal partners in American society. Some of the radical

In 1969, some California women held an Anti-Bra Day to protest the pressure that society put on women to wear constricting, "feminine" garments. "I ask college students not born in 1963, 'How many of you have ever worn a girdle?' They laugh," wrote Betty Friedan. "So then I say: 'Well, it used to be, not so long ago, that every woman from about the age of 12 to 92 who left her house in the morning encased her flesh in rigid plastic casing.' "

Portrait of Herself

Margaret Bourke-White was one of America's great photojournalists and a groundbreaker for women in the field. A photographer for both Life and the U.S. Air Force during World War II, she later went on assignments in India and Korea. In her autobiography, Portrait of Myself, published in 1963, Bourke-White wrote:

My father was an abnormally silent man. He was so absorbed in his own engineering work that he seldom talked to his children at all, but he would become communicative in the world of out-of-doors....If Father had

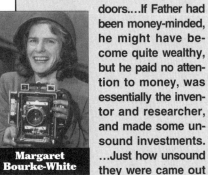

Margaret Bourke-White

been money-minded, he might have become quite wealthy, but he paid no attention to money, was essentially the inventor and researcher, and made some unsound investments. ...Just how unsound they were came out only when he died. I was 17 then, just starting college, and I know now that if we had been wealthy, and I hadn't had to work my way through college as I did after his death, I would never have been a photographer.... It is odd that photography was never one of my childhood hobbies when Father was so fond of it....Everything that had to do with the transmission and control of light interested my father, and I like to think that this keen attention to what light could do has influenced me.... That, and the love of the truth, which is requisite No. 1 for a photographer.

feminists wanted to overthrow society. They saw male-dominated institutions as hopeless. Robin Morgan wrote:

> I call myself a radical feminist....I believe that sexism is the root oppression, the one which, until and unless we uproot it, will continue to put forth the branches of racism, war, class hatred, ageism, competition, ecological disaster, and economic exploitation.

The radical feminists developed feminist publishing houses, health organizations, child-care centers, food cooperatives, and other women-run institutions. Some said they hated all men. Some wanted to integrate everything—even bathrooms. Although they were few in number, their extreme ideas got a lot of publicity.

By 1970, seven years after the publication of *The Feminine Mystique*, women activists were picketing and demonstrating for equal job opportunities and equal access to all-male clubs, restaurants, and schools. Feminists were making daily headlines in the newspaper—competing with the civil rights movement and the war in Vietnam.

The demonstrations made good newspaper copy, and they reflected a whole new attitude on the part of women. When one feminist leader was told, at a national political convention, to "calm down, *little girl*," she was anything but calm. All the brouhaha (BREW-ha-ha—it means uproar or hubbub) was making a difference. Some women began to get good jobs in

In 1966, 30 women, including Betty Friedan, founded the National Organization for Women (NOW), "to take action to bring American women into full participation in the mainstream of American society." In 1970, these women marched for equal rights up New York's Fifth Avenue.

banking, law, engineering, and other previously male-only fields. Women were now newspaper editors and TV anchors.

But that news was only part of the picture. Things did not go well for all women—especially poor, black, or Hispanic women. According to Chafe:

> From 1955 to 1981, women's actual earnings fell from 64 percent of men's to 59 percent, and even in the late 1980s their earnings had climbed back only to 62 percent—still below the figure thirty years earlier....Eighty percent of all women workers were employed in just 5 percent of all jobs—the lowest-paying 5 percent....By the end of the 1980s, one in every four children in America was poor, and women comprised almost 70 percent of the adult poor....Middle and upper-class white women might be experiencing a new freedom, but almost none of the benefits they derived from women's new opportunities trickled down to the poor. For these women, race, class, and gender represented a triple whammy.

The women's rights movement had hit a class and race wall. It was unintended. Betty Friedan and others had worked for racial as well as female equality. But it was the women who were educated and talented who were going places. If you were poor, it was difficult to get a really good education. (For those who did, no matter their race, there were jobs aplenty.) Women without training or skills—who made up a majority—were usually stuck in dead-end, low-paying jobs, and still suffering from sex discrimination. (Actually, the same thing was happening to men. Education was becoming more and more important for all as the 21st century approached.)

In 1977, thousands of women took part in a 2,610-mile marathon from Seneca Falls, New York—where Susan B. Anthony and Elizabeth Cady Stanton had drawn up a declaration that said "All men and women are created equal"—to Houston, Texas, to publicize the first National Women's Conference in support of an equal rights amendment to the Constitution. They were joined for the last lap by (starting second from left) tennis star Billie Jean King, Susan B. Anthony (great-niece of the first Susan), politician Bella Abzug, and Betty Friedan (far right).

Milestones of the Women's Movement

In 1920, Carrie Chapman Catt carried the torch for the 19th Amendment to the Constitution.

In 1791, Englishwoman MARY WOLLSTONECRAFT wrote: "It is time to restore women to their lost dignity and to make them part of the human race."

In 1821, EMMA WILLARD founded the Troy Female Seminary in Troy, New York, to give girls the kind of education that was available to boys.

Women who wear pants can thank AMELIA JENKS BLOOMER, who had the courage to get out of her hoopskirt—in the 1840s—and wear "bloomers." *New York Herald* editor James Gordon Bennett wrote, "If women mean to wear the pants, then they must also be ready in case of war to buckle on the sword!"

In 1848, ELIZABETH CADY STANTON led a Women's Rights Convention in Seneca Falls, New York, and issued a female Declaration of Independence that said "all men *and women* are created equal."

In 1857, when no hospital would let her practice, Dr. ELIZABETH BLACKWELL opened the New York Infirmary with an all-woman staff. Dr. Blackwell had never let adversity stop her. She was turned down by 29 medical schools before finally being accepted as a medical student.

In 1873, SUSAN B. ANTHONY was tried and convicted (by a court in Rochester, New York) of the crime of having voted!

In 1920, women finally got the right to vote when enough states ratified the 19th Amendment. Women suffragists had demonstrated (and suffered) to get it passed. CARRIE CHAPMAN CATT, a leader of the movement, then founded the National League of Women Voters.

There was something else, besides class and race. Women weren't all alike. They had different political ideas. Many women were *traditionalists*: they thought that a woman's primary role *was* as a wife and mother, and that a career detracted (took away) from that role. The traditionalists were conservative, but, like the radical feminists (who were not conservative), they emphasized the differences between men and women. Phyllis Schlafly (SHLAFF-lee) was the spokeswoman for the traditionalists. She wrote of the "Positive Woman" who "understands that men and women are different, and that those very differences provide the key to her success as a person and fulfillment as a woman." She attacked the women's liberation movement.

Schlafly was part of a powerful political force, the conservative "New Right," which developed in response to the turmoil of the times (the Vietnam war, growing rates of drug use, crime, and divorce, and new sexual ideas). The ideas of the New Right reached many Americans through something new: television church programs. TV preachers like Jerry Falwell, Pat Robertson, and Tammy and Jim Bakker were speaking to millions of Americans.

At the 1977 National Women's Conference, Phyllis Schlafly spoke against the equal rights amendment.

It was a Supreme Court decision that gave the New Right a focus. In 1973, in the case of *Roe* v. *Wade*, the Supreme Court said that a woman, in consultation with her doctor, could decide to end her pregnancy. That means she could choose to have an abortion. That decision gratified some Americans and outraged others.

Feminists said that women were finally in control of their own bodies. Anti-feminists said that abortion was murder and at odds with traditional religious values.

Does all this sound confusing? Well, change is rarely orderly. Women were sorting out new ideas. Men were, too. Fathers were enjoying their children and participating in home activities in ways that other generations had not done. And some of the statistics carried good news for women.

In the '50s and '60s, women had made up between 5 and 8 percent of the students in medical, law, and business schools. By the mid-'80s, they were at 40 percent, and heading upward. Educated women could now make choices. They could work as veterinarians or as housewives. They could be Supreme Court justices, brain surgeons, or, like that pioneer Lucille Ball, television producers. For women—of all races and backgrounds—there was now a hard question to consider: just how do you manage a job, marriage, children, a home, friends, and community involvement?

A Great Professional

In the 1950s, tennis was a country-club game. Mostly, it was a sport for the wealthy. Billie Jean Moffitt (who became Billie Jean King when she married) didn't belong to a country club. She learned to play tennis on public courts in California. When she entered her first tournaments, in 1955, she ran into some snobbish treatment. But King was a fighter. She fought her way to the top of the tennis world, she fought against elitism in tennis, she fought for equality for women in sports, and she fought for professionalism in tennis.

Billie Jean King

The big tournaments—for men and women—were open only to amateurs. That meant that champions played for the love of the game—not for money. (If you accepted money, you were thrown out of the tournament.) That was fine for those who were rich, but very hard on everybody else. By the 1960s, the U.S. Lawn Tennis Association was paying the top men and women players "under the table." King blasted that practice. She said it kept tennis from becoming professional. She called it "shamateurism." She said it wasn't fair. She wasn't alone.

In 1968, tennis became an open sport. Amateurism was finished. Anyone who qualified could compete for money prizes.

Now women players faced another problem. The prizes for the winners of men's tennis tournaments were as much as 12 times higher than the women's. The USLTA and most male tennis players said that women's tennis wasn't as good as men's tennis. They said that spectators weren't interested in watching women.

Billie Jean King said they were wrong. She and eight other top women players quit the USLTA–sponsored tournaments. They formed their own tour. They took a chance. They proved that women's tennis did attract a following. Crowds cheered the women on. Finally, the USLTA began to change its sexist practices.

In 1972, *Sports Illustrated* named Billie Jean King its first Sportswoman of the Year. When she retired, in 1984, King had won nine U.S. championships and a shelf full of other trophies. She'd helped bring respect and professionalism to women's athletics. (For a good book on the history of women's tennis, read *We Have Come a Long Way*, by Billie Jean King.)

30 As Important as the Cotton Gin

Do these houses in Savannah, Georgia, look like slave cabins? They were slums, torn down in 1941. By then, many southern blacks had given up and headed north for good.

Everyone knows about Eli Whitney's cotton gin, and that it made growing cotton profitable, and that cotton demands lots of fieldworkers, and how that led to the creation of an enormous slave society in the cotton-growing South.

But most people don't know what ended that cotton-growing society. It wasn't the Civil War. When slavery ended, most of the former slaves kept on picking cotton. Now they were sharecroppers, or tenant farmers, instead of slaves. Some of them could hardly tell the difference. They didn't own land, they lived in plantation-owned cabins, they picked for a plantation owner, they had almost no schooling, and they didn't have much control of their lives.

So what ended that cotton-picking life? A machine. The mechanical cotton picker. That machine could do the work of 50 people. When the mechanical cotton picker was demonstrated on a Mississippi farm in 1944, cotton sharecropping was on its way to extinction. Cotton workers were no longer needed—or wanted. If their cabins were torn down, cotton could be planted where they stood. The cotton pickers had to go.

In the old sharecropping days, blacks had sometimes been pulled off buses heading north. They were needed in the cotton fields. After 1945, some Mississippi communities began passing out free bus tickets for Chicago. Back in the 16th century, something like that had happened in England. Tenant farmers were thrown off the land because it was more profitable to graze sheep where they had farmed. The roads to London and Edinburgh became crowded with displaced farm workers and,

While the modern civil rights movement had a momentum of its own, the activism that characterizes the Afro-American community... was directly influenced by the generation of Afro-Americans who moved north during the Great Migration....After the Great Migration, American society was never again the same.

—SPENCER CREW, *FIELD TO FACTORY*

142

when those cities couldn't absorb them all, many kept going—right on to America.

Well, now the United States was seeing the same kind of exodus: from field to factory, from rural to urban, from South to North. This was an *internal*—inside the country—migration, but it was larger than that of any ethnic group to America from outside.

Between 1910 and 1970, 6.5 million blacks moved from the South to the North. Of that number, 5 million moved after 1940. They moved to Chicago, New York, Detroit, and Los Angeles. Others, who didn't go north, moved to the South's urban centers: Atlanta, Norfolk, Little Rock, Memphis, and Houston. A people with a tradition and culture based on farming became city folk. By 1970, more than three-fourths of black Americans lived in cities.

What was astonishing about all this was that hardly anyone in the rest of the country seemed to notice. Black people themselves didn't quite understand what was happening. Most Americans thought segregation was a southern problem. Its historical roots were southern. It was the legacy of slavery. The civil rights movement was something happening in the South, wasn't it? Why should anyone worry about racial problems in the North? They didn't exist, did they? Segregation in the North? Urban ghettoes? Poor schools? Poor jobs? The problems of the big cities? What problems?

You know about ostriches, don't you? They are the long-necked birds that supposedly hide their heads in the sand (they don't, really). Well, something big was happening in

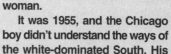

For Annie Chaplin of Beaufort, South Carolina, living conditions in the 1960s differed little from those of a century earlier. But was life in the North better?

As Good as a White Man

Emmett Louis Till was 14 when he was found, brutally beaten, shot in the head, wired to a heavy weight, and dumped into Mississippi's Tallahatchie River.

Why? Because he had dared to say a flirtatious word to a white woman.

It was 1955, and the Chicago boy didn't understand the ways of the white-dominated South. His mother had warned him when she put him on a train heading for Mississippi and his southern cousins. "If you have to get down on your knees and bow when a white person goes past, do it," said Mrs. Till.

Emmett Till

But that wasn't Emmett's way. He did just the opposite. Some boys dared him, and he went into a store, bought some candy, and then turned to the wife of the owner, Roy Bryant (who wasn't there), and said, "Bye, baby."

Within hours, two white men—Bryant and his brother-in-law, J. W. Milam—took Emmett away. Later, Milam told a writer exactly what had happened—how they beat and murdered him. Why did they do it? "Well, what else could I do?" said Milam. "He thought he was as good as any white man."

Bryant and Milam were found not guilty by an all-white jury (no blacks were registered to vote in the whole county, so there were no black jurors, either). Across the nation, all people of decency were outraged. "The murder of Emmett Till was the spark that set the civil rights movement on fire," wrote Sara Bullard in her book *Free at Last*.

Myrlie Evers, widow of civil-rights leader Medgar Evers (another murder victim), said that the Emmett Till killing showed that "even a child was not safe from racism and bigotry and death."

143

A War in Our Emotions

Timidly, we get off the train. We hug our suitcases, fearful of pickpockets, looking with unrestrained curiosity at the great big brick buildings. We are very reserved, for we have been warned not to act "green," that the city people can spot a "sucker" a mile away. Then we board our first Yankee street car....We pay the conductor our fare and look about apprehensively for a seat. We have been told that we can sit where we please, but we are still scared. We cannot shake off three hundred years of fear in three hours....Sometimes five or six of us live in a one-room kitchenette, a place where simple folk such as we should never be held captive. A war sets up in our emotions: one part of our feelings tells us that it is good to be in the city, that we have a chance at life here...that we no longer need bow and dodge at the sight of the Lords of the Land. Another part of our feelings tell us that, in terms of worry and strain, the cost of living in the kitchenettes is too high, that the city heaps too much responsibility upon us and gives too little security in return.

—RICHARD WRIGHT AND EDWIN ROSSKAM, *12 MILLION BLACK VOICES*

"I can save you the trouble right now," this real estate agent told a man looking for a house in a white neighborhood. "They just won't sell to you."

America—and the nation's leaders had their heads buried. All those people were moving into cities that were already crowded. They were being squeezed into places that weren't prepared. They needed good schools, they needed opportunity, they needed all the things America does so well—and it wasn't happening. The president, the Congress, the government experts, and the college professors weren't paying attention.

There were some exceptions. A few people noticed. One of them was a Mississippi lawyer, businessman, and writer named David Cohn. In 1947, he wrote:

> *The country is upon the brink of a process of change as great as any that has occurred since the Industrial Revolution....Five million people will be removed from the land within the next few years. They must go somewhere. But where? They must do something. But what? They must be housed. But where is the housing?...If tens of thousands of Southern Negroes descend upon communities totally unprepared for them psychologically and industrially, what will the effect be upon race relations in the United States? ...Will the victims of farm mechanization become the victims of race conflict?*
>
> *There is an enormous tragedy in the making unless the United States acts, and acts promptly.*

But the United States did not act. Most Americans remained unaware. All those people were soon packed into cities like powder in a firecracker. In the '60s, the fuses on the firecrackers were lit and the cities began to explode.

31 Picking and Picketing

Picking cotton in California. "We need Mexicans for their labor, for the same reason you need a mule," said one farmer.

Lucía and María Mendoza, 18 and 17, stumbled out of bed at 2 A.M., dressed in the dark, went into the kitchen of their adobe house, made a lunch of tacos and soda pop, and filled a thermos with hot soup. Then they woke their younger brother and their dad. Soon the four of them were heading north, toward the border between Mexico and the United States. They were on their way to pick lettuce in California. Each of them expected to make $16 that Tuesday. It wasn't much, but it was more than they could earn at home. They would use short-handled hoes that kept them bent over all day. It was hot, dusty, back-breaking work. The Mendozas were soon part of a line of cars making for the picking fields.

These Mexican farm workers were entering the United States legally. They were wanted to help harvest crops to feed people across the nation. But some of them would stay in the United States illegally. Hundreds of thousands of them had already done it. Most of those illegal immigrants had little schooling and few marketable skills. They were crowded

Mexico is sinking
California is on Fire
& we all are getting burned
  aren't we?
But what if suddenly the continent
  turned upside down?
what if the U.S. was Mexico?
what if 200,000 Anglo-Saxicans
were to cross the border
  each month
to work as gardeners, waiters,
3rd chair musicians, movie extras,
bouncers, babysitters, chauffeurs,
syndicated cartoonists, feather-
  weight boxers, fruit-pickers &
  anonymous poets?
what if they were called waspanos
waspitos, wasperos or waspbacks?
what if we were the top dogs?
what if yo were you
& tú fueras I, Mister?
—GUILLERMO GOMEZ-PEÑA, "MEXICO IS SINKING"

Picking peas, California, 1961. "We would start early, around 6:30 A.M., and work for four or five hours, then…eat and rest until about 3:30 in the afternoon," said one woman. "We would go back and work until we couldn't see."

Resent means to feel indignantly angry at someone.

in cities in poor districts (*barrios*). Their children needed to go to school. They needed job training and help. All that cost taxpayer money. Some Americans resented them.

Many said the Mexicans took jobs from American citizens, especially from Mexican-Americans. It wasn't their fault that they took those jobs. The growers wanted them instead of Americans because they would work for less money. They could live on less. Life in Mexico was cheaper than life in the United States.

Once they were across the border, the Mendozas parked their car and walked to a place where workers were hired for agricultural jobs. By 3:30 A.M. they were settled just behind the driver in an old, rattletrap bus heading north toward lettuce fields. Most of the 46 passengers tried to sleep. They knew they had a long ride ahead of them, and it was still dark. Later, one passenger would remember that they had been going very fast when the driver missed a curve and the bus became airborne—crashing into the bank of a canal, bouncing off that bank to the other bank, and settling in a shallow waterway. All the seats in the old

César Chávez speaks to migrant workers. They trusted him because he was one of them; he had lived their life and understood its hardships.

bus flew out of their sockets in a mess of arms, legs, twisted metal, and broken glass. Nineteen passengers were trapped in the bottom of the bus; they drowned in two and a half feet of water. The four Mendozas were among them.

César Chávez wept for Lucía and María and the others who lay in silent caskets. Chávez was an American of Mexican heritage. Like most of the 2,000 mourners at a special funeral mass for the victims, he was a devout Roman Catholic. Everyone knew Chávez. He was the leader of the Farm Workers Association, and famous. They knew he cared about people, especially farm workers. They wanted to hear what he had to say. Speaking in Spanish, Chávez told them:

> This tragedy happened because of the greed of the big growers who do not care about the safety of the workers and who expose them to grave dangers when they transport them in wheeled coffins to the field.
>
> The workers learned long ago that growers and labor contractors have too little regard for the value of any individual worker's life. The trucks and buses are old and unsafe. The fields are sprayed with poisons. The laws that do exist are not enforced. How long will it be before we take seriously the importance of the workers who harvest the food we eat?

Chávez knew all about harvesting food. He had been a migrant worker himself, traveling from bean fields to walnut groves to grape arbors, following the harvest of the seasons. That meant living in a tent, or whatever room could be found. When he was a boy, it meant changing schools as often as he changed picking fields. It meant sometimes not having shoes or a bathroom to use. By the time César graduated from eighth grade he had attended 38 different schools.

There was something special about Chávez, although it was hard to

Whenever the people are well informed they can be trusted with their own government.
—THOMAS JEFFERSON

Caskets are coffins.

This is the beginning of a social movement in fact and not in pronouncements. We seek our basic God-given rights as human beings. Because we have suffered—and are not afraid to suffer—in order to survive, we are ready to give up everything, even our lives, in our fight for social justice. We shall do it without violence because that is our destiny. To the ranchers, and to all those who oppose us, we say, in the words of Benito Juárez, EL RESPECTO AL DERECHO AJENO ES LA PAZ [which means *respect for human rights is peace*].
—CÉSAR CHAVEZ AND OTHERS, *THE PLAN OF DELANO*

147

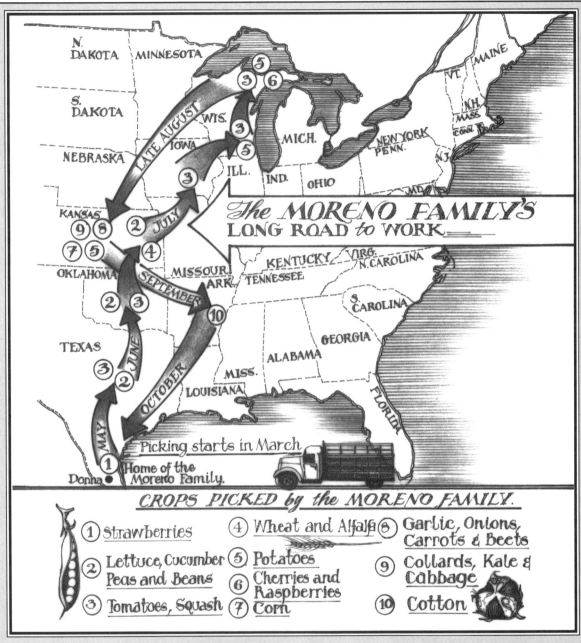

The MORENO FAMILY'S LONG ROAD to WORK

Picking starts in March

Home of the Moreno Family.

Donna ●

CROPS PICKED by the MORENO FAMILY.

① Strawberries

② Lettuce, Cucumber Peas and Beans

③ Tomatoes, Squash

④ Wheat and Alfalfa

⑤ Potatoes

⑥ Cherries and Raspberries

⑦ Corn

⑧ Garlic, Onions, Carrots & Beets

⑨ Collards, Kale & Cabbage

⑩ Cotton

Graciela Moreno grew up in Donna, Texas, near the Gulf of Mexico, in the 1950s. Until she was about ten years old, says Graciela, "In winter and spring I went to school. In the summer the whole family piled into a big flatbed truck and set off on a long, long trip." But it wasn't a vacation trip. The Morenos picked fruit and vegetables, starting with strawberries at home in March and moving slowly north as crops ripened. The kids picked alongside the grownups. "Sitting in the truck while we drove was boring," says Graciela. "But the camps where we ate and slept were kind of fun." They picked lettuces and cucumbers and peas and beans in Texas and Oklahoma and Kansas, and tomatoes and squash and potatoes in Iowa and Wisconsin. They picked cherries in Michigan and onions and carrots in Kansas again. They picked cotton in Arkansas. Back in Texas, they picked citrus fruit in October. It felt very good to come home again.

decide quite what it was. He had a pleasant, round face with brown skin and dark straight hair—there was nothing out of the ordinary about that. He was a gentle man, and he didn't boast or call attention to himself. But when he had a job to do he did it carefully and well. He could be trusted; he was honest, thoroughly honest. So when people needed help, they often turned to him.

Fred Ross, an organizer who came to California to try to help the farm workers, heard about César Chávez and gave him a job with the Community Service Organization, which helped poor people deal with many kinds of problems.

Chávez was soon helping those people find housing, medical care, food, and, if they needed legal aid, a lawyer. He got them to register to

Retired teacher Alice Barnes, CSO leader Fred Ross, and Chávez (left). Barnes, like many other ordinary citizens, marched again and again with the field workers.

vote, and he made them realize the power of the vote. Then he began to think about starting a labor union for farm workers. A labor union is an organization of people, usually all doing the same kind of work, who get together to try to make life better for themselves. Factory unions are easy to organize—most of the workers are together in one place—but getting agricultural workers organized isn't easy at all. In California, farm workers labored on thousands of farms that stretched the length of the state. Chávez knew that many growers took advantage of workers. They paid them little, they ignored unsafe conditions, they got their children to work even though that was against the law, and sometimes they cheated them on their pay. By themselves the workers had no power, but if Chávez could organize them into a union, they could demand fair wages and safe conditions.

César Chávez had a wife, eight children, and that steady job with the CSO. When he told his wife he wanted to quit his job to try to start a union, what do you think she said? She knew the family might go hungry if he had no regular work. What would you say? Helen Chávez said okay. She understood that if *La Causa*—the cause—were successful, it would help millions of people. And, knowing César, Helen Chávez thought there was a good chance it would be successful.

It was 1962, and César Chávez started going from farm to farm, talking to workers. Three years later, his Farm Workers Association voted to join Filipino farm workers in a strike against the grape growers. The workers refused to pick grapes until they got better pay and better working conditions. Then the growers hired other

Chávez in 1970, during the workers' strike against grape growers.

A California migrant workers' camp, 1935. This standard of living was normal for many until Chávez and the UFW fought for better pay and conditions.

Your ***conscience*** is your understanding of what is right and wrong; your knowledge of what is the right thing to do.

California's biggest industry is agriculture, and grapes are the biggest money crop. Before *La Causa*, those who picked the crops got little benefit from the riches they created.

pickers. Union members marched near the grape fields with signs that said ¡HUELGA!, which means *strike* in Spanish. Chávez convinced some of the new pickers to stop work and strike with them. Grapes began rotting because no one was picking them. The growers were furious; union members were attacked and beaten. The police helped the growers.

César Chávez had been inspired by Gandhi, Martin Luther King, Jr., and his own religious beliefs. He insisted that the farm workers fight with peaceful marches and prayers. Nonviolence, he told them, took more courage than violence. He also believed that it achieved more. It appealed to the conscience of good people everywhere.

Chávez needed to draw attention to *La Causa*. He decided that a 300-mile march across much of California might just do it. He got university students and religious leaders to agree to march with the farm workers. Look at a map and find someplace that is 300 miles from your home. Now imagine walking that far. Chávez's feet became blistered and his legs swollen. He could hardly walk—but he kept going. Television cameras whirred. Suddenly everyone knew about *La Causa*.

Some farm owners called César Chávez a communist (he wasn't), but most people believed he was on the side of justice and fairness. Finally, a few growers signed contracts with the union—but most still would not do it. (About this time, the Farm Workers Association was renamed the United Farm Workers; it became part of a national union—the AFL-CIO.)

Chávez announced a boycott. He was going to ask people across the United States not to buy grapes grown in California. But boycotts work slowly, and some of his union members were impatient. They wanted to use violent methods. Chávez had to do something to control them and to make the growers pay fair wages. He did what Gandhi did. He went on a fast. For 25 days he ate no food. Finally, 26 growers signed contracts with the union.

César Chávez started eating again.

Farm workers struck for the right to unemployment and health insurance as well as decent wages.

32 "These Are the Times That Try Men's Souls"

Poor People's Campaign 1968

"We will place the problems of the poor at the seat of the government of the wealthiest nation in the history of mankind," said Dr. Martin Luther King, Jr.

That astonishing idea—that all people are created equal and are equally entitled to pursue happiness—still seemed revolutionary in the 1960s, almost 200 years after the days of Thomas Jefferson and Tom Paine. All men and women did not have equal opportunity in the United States in the 20th century: some were privileged and some were disadvantaged. The American Revolution was unfinished.

Yet there had been enormous changes in the two decades since the end of World War II. Martin Luther King, Jr., could look at the South with some satisfaction. Where segregation had once flourished, blacks and whites now worked together, voted together, went to school together, ate in restaurants together, and rode on buses together. The changes were amazing to those who knew the old South. The new racial harmony had helped bring industry and a progressive spirit to the South. The region was thriving.

But when King decided to take his movement north—into big-city ghettoes—he discovered problems that were tougher than any he had faced before. He had believed that the methods that worked in the South would work in other regions, too. But the situation was different in America's northern cities.

In the South, the problem had been Jim Crow laws and police-enforced segregation. The solution was to get the laws changed and to get the police to enforce those laws. The job wasn't finished—but the laws were in place and the direction was clear.

There were no Jim Crow laws in the North or West. The urban nightmare had nothing to do with laws. The problem there was economic.

"These are the times that try men's souls," wrote firebrand Tom Paine of the violent revolutionary days of 1776.

How many years in a decade?

Flourish means to grow and thrive.

A ***ghetto*** is a city neighborhood where poor people live crowded together, usually in bad conditions.

151

The Bones of Their Ancestors

The old woman was wheeled into the Senate. No one knew how old she was; even she didn't know. Maybe in her eighties. She had come a long way, flying inside a steel bird across a land that had once belonged exclusively to birds and fish and animals, and then to her ancestors (who had intruded from Asia). Now it was called home by a rainbow of humans (and by robins and salmon and grizzly bears, too). Mary Jim Chapman came from the state of Washington (which was lapped by the waters of the Pacific) to the city of Washington (which got some of its breezes from the Atlantic). She came from the Yakima Indian Reservation (with boundaries that fenced in once-nomadic Indians) to a place where the great chiefs of the land sat in buildings of marble and granite.

Her daughter Carrie pushed her chair into the Senate hearing room, and Senator Daniel Inouye asked her to speak. Mary Jim cleared her throat, and in high, trembling tones, sang out: *Ayyyaaaaaaaaaaaa*. It was the wail of a people. It got attention.

She had come for what belonged to her and the other Palouse: the bones of their ancestors. They had been wrested from their graves on Fishhook Island by archaeologists who claimed scientific rights. That was in 1959. It was now 1988; Mary Jim had spent all those years obsessed, haunted, miserable, feeling incomplete. Sociolo-

gists explained this. They said that the Native American cultures were different from the other culture—the one called "Western," or "dominant." For Native Americans, life was a woven ribbon that was centered in community and continuity, they explained. One generation was responsible for both those who came before and those who were to come. To disrupt that pattern was to tear the fabric.

But perhaps the differences weren't as great as the sociologists believed. How would those sociologists feel if the bones of their mothers and fathers were ripped from the earth and put in museums for strangers to touch and chuckle over?

Mary Jim Chapman and her daughter went to the Smithsonian Institution's National Museum of Natural History. There were Indian bones there. They were not Palouse bones from Fishhook Island. So they went home, still haunted by their obligation to their ancestors.

And then in 1991, Roderick Sprague, an archaeologist from the University of Idaho, pulled into their driveway. He had dug Indian bones when he was younger. Now he too was obsessed with returning the bones to their homes. Finally, 34 skeletons were found and reburied on Fishhook Island. But the skeletons did not include Chapman's grandfather, Chow-wah-what-yuk, who had been buried in his canoe. The search for his remains continues.

The cities were filled with poor people—black, white, brown—who weren't being given a chance to rise out of poverty. Usually they went to schools that were tattered and poor, where they didn't get good training for the new kinds of jobs that technology was bringing. But there were hardly any jobs in the inner cities, anyway. Many city people were almost without hope. Young blacks in the cities were full of frustration and rage.

And urban whites? When King marched in the Chicago area he was met with white hatred more vicious than anything he had encountered in Mississippi or Alabama. Many of the city's whites were poor, too. They were competing with blacks for jobs—and there weren't enough jobs to go around. Instead of coming together, black people and white people seemed to be growing farther and farther apart. Some leaders, on both sides, were encouraging hatred.

The war in Vietnam wasn't helping. The soldiers who came home had been trained to be violent and wield weapons. Many had learned to use drugs in southeast Asia. They were like lighted matches in those packed cities.

President Johnson still hadn't found a way to get out of Vietnam. He was putting pressure on those, like Martin Luther King, Jr., who protested against the war.

It was a tough time for King. John Kennedy's brother, Robert (Bobby) Kennedy, was telling him he needed to bring his battle for justice north. Kennedy said the same thing to the ministers and the others who had gone south. To the thousands who had marched from Selma to Montgomery, he said:

But the brutalities of the North receive no such attention. I have been in tenements in Harlem in the past several weeks where the smell of rats was so strong that it was difficult to stay there for five minutes, and where children slept with lights turned on their feet to discourage [rat] attacks. Thousands do not flock to Harlem to protest these conditions.

Martin Luther King, Jr., decided to begin a new campaign. It would be a campaign against poverty. King's program was aimed at "all the poor, including the two-thirds of them who are white." Poverty was not just a black problem, or just a white problem—it was a national disgrace. King planned to bring poor people to Washington. This would not be a one-day march. They would stay; they would camp in the city; the government leaders would have to pay attention.

King believed that eliminating poverty made economic sense, as well as being the right thing to do. If you turn the poor into purchasers, King said, they will solve many of their own problems. The vast amounts of money that the nation spent on military goods could be used to make life better for all people. "Poverty has no justification in our age," he said. "War is obsolete." He believed that nonviolence was the way nations must learn to solve their differences. But time was running out for King and his nonviolent ideas. "The only time that I have been booed," said Dr. King, "was one night in a Chicago mass meeting by some young members of the Black Power movement."

King talked over his problems with Bobby Kennedy. So did César Chávez. Kennedy really seemed to care about poor people. Few other politicians did. The senator was now on the campaign trail, attempting to win the Democratic nomination for president. It looked as if he might do it.

Coretta Scott King (outside the gate, center) demonstrates against the war in Vietnam, which many civil rights activists saw as a futile distraction that diverted the government's money and attention away from terrible problems at home.

Bobby Kennedy campaigns for the Senate in New York. "All of us in all sections of the country have a great lesson to learn," he said. "The importance of getting a dialogue going between people in the North and South."

Some statistics:

In 1991 alone, there were 38,317 firearm deaths (murder, suicide, accidents) in the United States. Compare that with 47,364 American war deaths during all the years of U.S. involvement in Vietnam.

In 1992, handguns killed 13,220 people in the U.S.; 36 in Sweden; 60 in Japan; 128 in Canada; 33 in Great Britain; and 13 in Australia.

Civil Rights for Native Americans

The civil rights movement inspired Indian leaders. They, too, had faced discrimination and persecution. They knew that many who were Indian pretended to be otherwise because they were ashamed of their heritage. But that was changing.

In the state of Washington (Mary Jim Chapman's home), tribes held "fish-ins" to protest restrictions on their treaty-given right to harvest salmon. In Minneapolis, urban Indians created AIM (the American Indian Movement) and shouted of Red Power. In San Francisco, like-minded Indians formed Indians Of All Tribes, and when the Bureau of Indian Affairs refused to listen to their grievances, they seized an abandoned federal prison on Alcatraz Island in San Francisco Bay and held it for a year and a half (much of it spent under the glow of TV klieg lights).

The Indians occupying Alcatraz offered to pay $24 in beads and cloth for it, the price paid for Manhattan 300 years earlier.

Traditionally, Indians had focused on their tribal identity. Alcatraz was a *pan-Indian* action. That means that Indians from tribes that had sometimes been one another's enemy were now banding together—although it didn't happen easily. Many tribes, especially those on reservations, were still determined to go their own way.

The militant activists went from Alcatraz to the Bureau of Indian Affairs in Washington, where they staged a protest occupation in 1972. Next they took over a trading post at Wounded Knee, South Dakota, where there had been a terrible massacre of Indians in 1890.

Those very dramatic takeovers accomplished what was intended. They made non-Indians aware of Indian grievances. And they made some Indians rediscover and take new pride in their heritage. Mohawk Richard Oakes said that Alcatraz was not "a movement to liberate the island, but to liberate ourselves."

The most tangible successes came to Native Americans through the courts. Again and again the courts decided in their favor in disputes over rights granted in treaties (many signed 100 or more years earlier).

In 1971, Aleuts, Eskimos, and other native Alaskans won 40 million acres of land and nearly $1 billion in settlements of long-standing claims. In Maine, Penobscots received $81 million for claims based on a law passed in 1790. Several other tribes received similar awards.

Overall, however, things were still not good for many Native Americans. Alcoholism devastated whole peoples. Unemployment was high on reservations and among urban Indians.

But some Indian entrepreneurs were taking a new look at Indian reservations, which were nations within a nation and thus not subject to most state requirements. Indian businesspeople realized that the status of reservations allowed for activities that were often illegal outside, like gambling. In the 1980s and '90s, casinos began to bring enormous wealth—which meant jobs, good schools, and nice homes (along with controversy and power)—to some Indian reservations.

Other things were happening: in Washington, D.C., the Smithsonian Institution returned Indian skeletal remains and funeral objects to their rightful owners. In Colorado, voters elected an American Indian, Ben Nighthorse Campbell, as a representative to the U.S. Congress (and as a senator in 1992). And in Oklahoma, Wilma Mankiller, the first female chief of the Cherokee Nation, spoke of her "firm belief that 500 years from now there will be strong tribal communities of native people in the Americas where ancient languages, ceremonies, and songs will be heard." If so, all the peoples of the land will be richer for it.

33 Up to the Mountain

In a speech after the Selma march, Dr. King said, "I know you are asking, 'How long will it take?' I come to say to you...it will not be long, because truth pressed to earth will rise again."

Martin Luther King, Jr., was preparing for the Poor People's Campaign in Washington when the garbage workers in Memphis, Tennessee, went on strike. They needed help, and King agreed to lead a march on their behalf. That march had hardly begun—King was in the front row—when teenagers at the back of the line began smashing windows and looting stores.

King was furious. "I will never lead a violent march," he said. "Call it off." A staff member urged the marchers to turn around and return to the church where they had begun. Dr. King left. But the police and the rock-throwing youths weren't finished. By the time they were, 155 stores were damaged, 60 people were hurt, and a 16-year-old boy had been killed by police gunfire. It was the first time that anyone had been killed in a march led by Martin Luther King, Jr. He felt sick that a boy had died. He was horrified by the violence. He couldn't sleep. What should he do? he asked a friend. "It may be that those of us who [believe in] nonviolence should just step aside and let the violent forces run their course, which will be...very brief, because you can't conduct a violent campaign in this country."

But King couldn't step aside. He decided that he had to lead a peaceful march in Memphis. "We must come back," he said, "Nonviolence...is now on trial." Some of Dr. King's

Memphis's black garbage workers formed a union and went on strike (above). They struck because some black workers—but no white workers—had been sent home one day when it rained. When the rain stopped, the whites went back to work and were paid a full day's wages; but the blacks, because they had been sent home, were paid for only a few hours.

155

"Tonight I want to speak to you of peace in Vietnam and southeast Asia," said LBJ in the speech with which he announced that he would not run for president again. "In the hope that this action will lead to early talks, I am taking the first step to de-escalate the conflict. We are reducing the present level of hostilities. And we are doing so unilaterally, and at once."

What does de-escalate *mean?* Hostilities? Unilaterally?

aides didn't agree. They thought Memphis was too dangerous. J. Edgar Hoover, the head of the FBI (the country's federal law-enforcement agency), hated Dr. King. He was using illegal methods to tap King's phone, and he was starting rumors and planting false articles in newspapers. Later, the truth came out about Hoover, but right now Dr. King was receiving death threats in the mail. That didn't stop him. He was going to go back to Memphis.

The night before his trip, King turned on the television. President Johnson was making an announcement. First Johnson said that he was cutting back on the bombing of North Vietnam and would try to get a settlement of the war. That was a surprise—and a relief. Then Lyndon Johnson stunned the nation. "I shall not seek and I will not accept the nomination of my party for a second term as your president," he said. The big man who wanted to be the greatest of all presidents, who wanted to end poverty, who wanted to do his best for America, had failed. The war had claimed another victim.

The very next evening, in Memphis, Dr. King spoke before a huge crowd at a church rally. He didn't have a written speech; he just spoke from his heart. He pretended that he was at the beginning of time and God was asking him, "Martin Luther King, which age would you like to live in?" Among the times he considered was when Moses led the children of Israel out of slavery in Egypt. Then he wondered about the time when the gods of the ancient Greeks were believed to live on Mount Olympus. He imagined what it would be like to see Martin Luther nail his 95 arguments to the church door in 16th-century Germany. He thought about being with Lincoln in 1863, when the president signed the Emancipation Proclamation. He even considered the time of Franklin Roosevelt and the problems of worldwide war. But King decided that none of those were the times he'd choose. "Strangely enough," he said, "I would turn to the Almighty and say if you allow me to live just a few years in the second half of the 20th century, I will be happy."

Now that did seem a strange choice to make, but King said, "Only when it is dark enough can you see the stars." The civil rights leader understood that he and many others of the 20th century were grappling with problems of the first order: war and peace and human rights. Everywhere, people were rising up, saying, "We want to be free." Was there any time in history that was more important? Those who heard him that evening would always remember his next words:

I would like to live a long life. But I'm not concerned about that now. I just want to do God's will. And He's allowed me to go up to the mountain. And I've looked over. And I've seen the Promised Land. And I may not get there with you. But I want you to know tonight that we as a people will get to the Promised Land....I have a dream this afternoon that the brotherhood of man will become a reality.

The next evening, after making plans for the Memphis march, Martin Luther King, Jr., went out onto the balcony off his room at the Lorraine Motel to breathe some fresh air before dinner. His friend Ralph Abernathy heard something that sounded like a firecracker. But it was no firecracker. Martin Luther King, Jr., had been shot dead.

Robert Kennedy heard the news in Indianapolis, just before he was to speak to a black crowd in a troubled section of the city. The people on the street had not heard the awful news. "Cancel the talk," the mayor of Indianapolis urged. The police refused to protect the senator. But Kennedy would not leave. He climbed onto the flat back of a truck under some oak trees and told the crowd of the tragedy in Memphis. Then he said:

Martin Luther King dedicated his life to love and to justice for his fellow human beings, and he died because of that effort. In this difficult day, in this difficult time for the United States, it is perhaps well to ask

Someone in the Hebrew Bible (the Old Testament) led his people to the mountaintop and looked over into the Promised Land. Who was it?

Senator Robert Kennedy said he supported a "massive effort to create new jobs—an effort that we know is the only real solution." A friend said that he "would have torn the country apart to provide jobs for everybody."

It is April 4, 1968, and on the balcony at Memphis's Lorraine Motel, Martin Luther King's companions point to the source of the shots from the high-powered rifle that killed the civil rights leader. King was 39 years old. The assassin, James Earl Ray, was captured two months later at London Airport, in England.

Some think that Martin Luther King, Jr.'s last speech, the night before he was killed, showed that he had a premonition that he would die. "So I'm happy tonight," he said. "I'm not worried about anything. I'm not fearing any man. 'Mine eyes have seen the glory of the coming of the Lord.' "

When he shall die,
Take him and cut him
out into little stars,
And he shall make the
face of heaven so fine
That all the world will
be in love with night,
And pay no worship to
the garish sun.

Those are Shakespeare's words (from Romeo and Juliet*). Bobby Kennedy said them in Indianapolis the night Martin Luther King, Jr., died.*

what kind of a nation we are and what direction we want to move in. For those of you who are black—considering the evidence there evidently is that there were white people who were responsible—you can be filled with bitterness, with hatred, and a desire for revenge. We can move in that direction as a country…black people amongst black, white people amongst white, filled with hatred toward one another.

Or we can make an effort, as Martin Luther King did, to understand and to comprehend, and to replace that violence, that stain of bloodshed that has spread across our land, with an effort to understand with compassion and love.

He told his listeners that he understood their anguish because he had lost a brother to an assassin's bullet.

What we need in the United States is not division; what we need in the United States is not hatred; what we need in the United States is not violence or lawlessness, but love and wisdom, and compassion toward one another, and a feeling of justice towards those who still suffer within our country, whether they be white or they be black.…The vast majority of white people and the vast majority of black people in this country want to live together, want to improve the quality of our life, and want justice for all human beings who abide in our land.

Let us dedicate ourselves to what the Greeks wrote so many years ago: to tame the savageness of man and to make gentle the life of this world.

Let us dedicate ourselves to that, and say a prayer for our country and for our people.

The crowd was hushed; people wept; and there was no violence.

ALL THE PEOPLE

34 A New Kind of Power

Martin Luther King, Jr.'s final journey to Atlanta, in a mule-drawn farm cart, was broadcast by satellite to millions all over the world.

Martin Luther King, Jr., was carried to his grave in a casket of polished African mahogany on a plain farm cart pulled by two mules. The cart and the mules reminded people that King's ancestors had farmed America's land with courage and dignity. The mahogany symbolized his African heritage. Weeping at the graveside were leaders from around the world, who had come to pay tribute to the man who had earned a Nobel Peace Prize with his message of love and brotherhood and peace.

But, at the very time King was being lowered into the ground, 130 cities around the nation were burning.

> **Each time a man stands up for an ideal, or acts to improve the lot of others, or strikes out against injustice, he sends a tiny ripple of hope, and crossing each other from a million different centers of energy and daring, these ripples will build a current which can sweep down the mightiest walls of oppression and resistance.**
>
> —ROBERT KENNEDY, SPEAKING IN SOUTH AFRICA

Rioters—looting and shooting—were killing people and destroying homes and businesses; 65,000 troops had to be called in to put down the riots. Almost all the victims were black. It didn't make sense. "We are living in an era when the lunatics, not the leaders, are writing history," wrote columnist Mary McGrory in the *Washington Star*.

When the fires cooled, 39 people were dead. The rioters said they were responding to the murder of Martin Luther King, Jr. But was that

Riots erupted all over the country in the wake of Martin Luther King's death. In Washington, D.C., seven people were killed, and over 1,000 wounded. At the Democratic National Convention in Chicago (above), anti-war demonstrators—even some delegates and newsmen—were clubbed by police as TV cameras rolled. Students and Black Panther activists staged anti-war protests and takeovers at colleges (below, at Cornell University).

History will say that my voice—which disturbed the white man's smugness, his arrogance, and his complacency—that my voice helped to save America from a grave, possibly fatal catastrophe.

—MALCOLM X

the right thing to do in memory of a man who had dedicated his life to nonviolence? Hadn't they heard his message?

Most black people had. Every poll showed that the majority of African Americans approved of the ideas of Martin Luther King, Jr., and disapproved of violence. But a black minority—a strong, active minority—was listening to other voices. Mostly those voices were young, male, urban, and angry. They were Black Power leaders; they wanted to change their world, and it certainly needed changing.

Some of them seemed to want power so they could get even for the terrible oppression of slavery and segregation. Some, disgusted by all oppression, wanted to separate themselves from whites. But some others wanted to bring respect and power to a black community that could then act on equal terms with whites.

That first idea didn't go far. Most black people had no intention of being oppressors. A few did want to separate themselves from the rest of America's citizens, which, after the sacrifices of the civil rights time, was difficult for many to understand. But that idea of power through respect—now that was appealing. Soon blacks—and whites, too—were studying African-American history. They were also learning about Africa and its history. They were wearing African-inspired clothes. They were telling stories of slavery from the slaves' point of view. They were taking pride in an inheritance full of stories of achievement. They were voting and electing blacks as sheriffs and mayors and congresspeople.

Black writers were bringing new sensitivities to readers. They were not just writing for African Americans; they were writing for all people. In 1940, Richard Wright published *Native Son;* five years later, his *Black Boy* was a main selection of the Book of the Month Club (both are books I recommend). Ralph Ellison had much of America tied up in his genes. His an-

Richard Wright

cestry was black, white, and Native American. In *Invisible Man* (1952), he wrote of the ways in which society ignores the ordinary person and makes him feel invisible and powerless and sometimes less than whole. In a stunning first novel titled *Go Tell It on the Mountain* (1953), James Baldwin wrote about the religious awakening of a boy living in Harlem.

Black women were among the best writers of the time. Zora Neale Hurston (who was part of the before–World War II Harlem Renaissance) was rediscovered and celebrated. Hurston's great novel *Their Eyes Were Watching God*—which is both funny and profound—inspired many other writers. Toni Morrison was one of them. She won the Nobel prize for literature—there is no higher honor. Alice Walker, Maya Angelou, and Paule Marshall, too, found power in words and ideas.

> To create one nation has proved to be a hideously difficult task; there is certainly no need now to create two, one black and one white.
> —JAMES BALDWIN

Alice Walker

Toni Morrison

Maya Angelou

Malcolm X

Malcolm X found power as a speechmaker. Malcolm had quit school, become a thief and a drug peddler, and landed in jail. He was frustrated; he wanted to turn his life around. But he couldn't express himself because he didn't have control of the English language. He decided to do something about that. He got a dictionary from the prison school and carefully copied every word onto a tablet. "With every succeeding page, I also learned of people and places and events from history." As his vocabulary grew, so did his sense of power and confidence. He began to read:

Anyone who has read a great deal can imagine the new world that opened up. Let me tell you something: from then until I left that prison, in every free moment I had, if I was not reading in the library, I was reading in my bunk. You couldn't have gotten me out of books with a wedge....Months passed without my even thinking about being imprisoned. In fact, up to then, I had never been so truly free in my life.

> My dearest friends have come to include all kinds—some Christians, Jews, Buddhists, Hindus, agnostics, and even atheists! I have friends who are called capitalists, socialists, and communists! Some of my friends are moderates, conservatives, extremists—some are even Uncle Toms! My friends today are black, brown, red, yellow, and white!
> —MALCOLM X

> ***The total number*** of white people who were poor at the end of the 1960s was larger than the *total number* of black or Hispanic people in poverty—but the *percentage* of blacks, other minorities, and especially women who were poor was very high.

At the 1968 Olympic Games in Mexico City, Tommy Smith and John Carlos—the gold- and bronze-medal winners in the 200-meter race—raised fists in the Black Power salute as the American national anthem was played, and refused to look at the flag. They were suspended from the games.

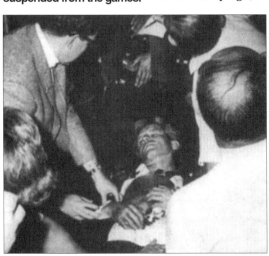

Two months after Dr. King's death, Robert Kennedy was killed by Sirhan Sirhan, who was born in Jerusalem and hated Kennedy for his support of Israel.

In 1960 blacks had very little political power. In that year there were only a few more than 100 black elected officials in the whole United States; by 1993 there were more than 8,000, including 40 members of Congress. Thurgood Marshall and Clarence Thomas had been appointed justices to the Supreme Court. Black people were millionaires (especially in the worlds of entertainment and sports).

Between 1950 and 1990, the number of African Americans in white-collar jobs—which means those who work in offices—leapt from 10 percent to 40 percent of all black workers. African-American men and women were engineers, doctors, lawyers, politicians, ballplayers, government workers, and artists. Many had good jobs, lived in beautiful houses, and belonged to fancy clubs.

Although most people still seemed to think in racial terms, that concern was just obscuring (which means hiding) the real problem. It was poverty in this prosperous land. Martin Luther King, Jr., had seen that. Bobby Kennedy understood that America would never truly be a land of the free if some people were trapped in poverty and inequality. "Today, in America," he said, "we are two worlds." They were the worlds of rich and poor. He said he hoped to build a bridge between those worlds.

Kennedy decided he would run for president; there were many who believed he would win that prize. And so he set out, giving speeches across the country. Young people flocked to his side. "It was an uproarious campaign, filled with enthusiasm and fun," his biographer wrote. Yet some, who had hated his brother, hated Bobby too. Wherever he went, along with the cheers there were also hate pamphlets. But in California, two months after Dr. King's funeral, there was cause for celebration. Bobby had won the Democratic primaries in California and South Dakota. "Here is [California], the most urban state of any of the states of our Union, South Dakota the most rural of any of the states of our Union. We were able to win them both. I think that we can end the divisions within the United States." On June 5, 1968, in front of a cheering crowd, he thanked some of those who had helped him: his staff, his friends, his wife, and César Chávez.

Then Robert Kennedy, heading for a press conference, took a short cut through the hotel kitchen. A shot rang out—and the man who might have been president was no more. It was the end of an era.

Later, a historian wrote, "Born the son of wealth, he died a champion of outcasts of the world."

35 The Counter Culture Rocks

Giving the peace sign at an anti-war protest. "Peace, man," became a way of saying hello.

In the '60s, a group of young people—mostly college age and middle class—started living differently from most Americans. They wore different clothes; they marched; they demanded power in their schools and colleges; sometimes they went off to live in their own little communities, called *communes*; and some refused to serve in the army because they didn't believe in fighting.

They were part of something that was called the *counterculture*. It had nothing to do with counters, and a few people (who felt threatened by those who weren't in the mainstream) said it had nothing to do with culture, either. But according to the dictionary, culture is *behavior patterns, arts, beliefs, institutions, and all other products of human work and thought characteristic of a population.* And that was what the counterculture was all about: behavior patterns. People in the counterculture just didn't behave as most other people did in the 1960s.

One meaning of the word *counter* is "against," and those in the counterculture stood against many of the ideas that guided the Vietnam era. Some people called them *hippies*; some people called them strange.

Mostly, they were energetic and idealistic. They were Jewish and Catholic and Protestant and Muslim and Buddhist. They had skin tones that were chocolate and honey and peach and mustard. They were male and female. None of that seemed to matter. What did matter was music and protest and ideas.

They thought the Vietnam War was wrong and immoral. They

Someone who won't serve with the military because of deep religious convictions (his conscience tells him or her that it is wrong to fight) is called a *conscientious objector*.

Many Americans saw communes only as havens for drugs and free love, but they were part of a long American tradition: the utopian community that tries to realize ideals of sharing and cooperation.

163

A national scandal erupted in 1970 when students at Ohio's Kent State University protested against our invasion of Cambodia. The National Guard was called in, and the nervous guardsmen opened fire. When it was over, four students were dead and ten hurt.

were civil rights marchers and they helped register new voters. Many wore their hair long and their clothes loose and colorful. Many lived in California, and they made other Americans realize that much of the nation's population had shifted west—and that maybe its ideas had shifted, too. They had big dreams: they wanted to make America live up to its ideals, and they might have achieved more, if it hadn't been for some of their experiments, like drugs, which turned to disaster.

But they did change things. They questioned everything, refused to conform, and made their favorite music—rock music—a national passion. Rock was a throbbing, pulsing, new kind of sound that took advantage of the electronic wizardry that was just being developed. It was urban music. It merged sounds from the music of blacks and whites. It was speeded up, and loud, and had a beat that was repeated and repeated so you couldn't get it out of your head. Some of it was political music. Some of it was disturbing music.

Rock wouldn't have happened, at least not quite the way it did, if it hadn't been for an unusual man—a very rich white man—who was a descendant of the 19th-century railroad tycoon Cornelius Vanderbilt. His name was John Henry Hammond, Sr., and he was born in 1910; but he might have been a child of the '60s, he was so far ahead of his times. Hammond grew up in a New York mansion on East 92nd Street with two elevators, a ballroom that could seat 200 people, and 16 servants. The African-American servants had a phonograph—an early record play-

Drugs and the Sixties

The drug culture was one part of the counterculture, and it did untold harm to many—those who experimented with drugs at the time, and thousands who came after them. The '60s produced the first generation in America to make taking drugs fashionable among a large group in society. But drugs take a terrible toll. Among the well-known musicians, actors, and entertainers of the '60s, '70s, and early '80s who died of drug overdoses (or related effects) were:

JOHN BELUSHI, actor

JOHN BONHAM, musician (Led Zeppelin)

LENNY BRUCE, comedian

TIM BUCKLEY, singer

JUDY GARLAND, actress and singer

ANDY GIBB, singer

TIM HARDIN, singer

Jimi Hendrix

JIMI HENDRIX, musician

BRIAN JONES, musician (the Rolling Stones)

JANIS JOPLIN, singer

FRANKIE LYMON, singer (the Teenagers)

KEITH MOON, musician (the Who)

JIM MORRISON, musician (the Doors)

RICKY NELSON, singer

PIGPEN, musician (the Grateful Dead)

ELVIS PRESLEY, singer

SID VICIOUS, musician (the Sex Pistols)

Janis Joplin

John Hammond

er—and John learned to love the music they played.

He had so much money that he could do whatever he wanted, and what he wanted was to listen to black music (especially jazz), and learn about black culture, and make recordings of the music that was now becoming his too. He dropped out of Yale College and drove around the country in a convertible in search of black musical talent, and he put that talent on record. Because of John Hammond, a whole young generation could turn on the radio and hear music that had African ancestry.

One of Hammond's first great discoveries was a jazz singer named Billie Holiday. Billie Holiday was a woman, and black. Hammond convinced white bandleader Benny Goodman to make records with her. Goodman was doubtful; blacks and whites didn't perform together in the well-known bands. But when Goodman heard Billie Holiday, he forgot his doubts. Then Teddy Wilson joined the Benny Goodman Trio and became the first black musician in a big-time white American jazz band. That bringing together of black and white musicians led to a mixing of white folk and country music with black rhythm and blues, jazz, and gospel (church) music. (All this happened at about the same time that radio and records became really popular, which was before World War II.)

Billie Holiday

Hammond made lots of other discoveries, and, in the late '30s, he staged a concert where an integrated New York audience listened to black music—from spirituals to jazz. Most of the audience had never heard music like that before; the concert was a sensation. Hammond not only liked black music; he also liked the people who made it. He wrote about black issues for magazines and served as a vice president of the NAACP (the National Association for the Advancement of Colored People). But he understood that talent has no color, and he discovered some terrific white musicians, too.

In 1961 he found a 20-year-old songwriter and performer in Minnesota who called himself Bob Dylan (although his real name was Robert Allen Zimmerman). Hammond got a big record company to sign Dylan to a contract. Dylan became the most influential musician of the Vietnam era. He was a poet who played a guitar and a harmonica and wrote music and

With rock and pop came discothèques, miniskirts, and dances: the twist, the mashed potato, the shake, the swim, the locomotion.

The importance of John Hammond to jazz history cannot be overestimated....Fortunately, his taste in jazz was impeccable. One after another, he brought forward young musicians who turned out to be important players; he recorded them, got them signed to good managers, and found them work. Among the people whose careers he furthered are Bessie Smith (he produced her last record session and paid the musicians out of his own pocket), Billie Holiday, Count Basie, and Charlie Christian.

—JAMES LINCOLN COLLIER,
*THE MAKING OF JAZZ:
A COMPREHENSIVE HISTORY*

"How many ears must one man have / Before he can hear people cry? / How many deaths will it take till he knows / That too many people have died?" sang Bob Dylan (above). When the Beatles sang, "I wanna hold your hand," millions of teenage girls wept, screamed, and fainted.

John Hammond
was an adventurous listener; he liked many kinds of music. He had courage, too. It took some courage when he signed folk singer Pete Seeger to a contract. Seeger had protested against the Vietnam War and was under indictment for contempt of Congress. Hammond also signed the Four Tops, a great Motown-sound group of the '60s—and in 1973 he signed a songwriter-guitarist named Bruce Springsteen (who went on to become one of rock's all-time superstars).

lyrics about the worries of the times. He was against the war and for civil rights, and his songs "Blowin' in the Wind" and "The Times They Are A-Changin' " became theme songs for the counterculture.

Dylan was intense and intellectual and sad. He wrote his own words and music, and his lyrics had a message. Everyone sang Dylan's songs, including Elvis Presley. The Beatles said that Dylan was their hero. Those four English musicians weren't sad at all; they were charming and inventive and electric—full of energy and terrific tunes. And talk about making money—the money they brought to England reduced its foreign debt. The queen gave them all medals (later, John Lennon gave his back when he disapproved of things the British government was doing).

Some said there was a kind of rivalry between Dylan and the Beatles, but that wasn't true. Beatle Paul McCartney said, "It was really a question of everyone admiring Dylan—and we felt kind of honored that he admired us." But it was the Beatles who became the most important popular musicians of their time (maybe of all time).

The Beatles had listened to a lot of black music on the radio in Liverpool, England, the big industrial port city where they grew up. They loved its drum beat, its rhythm, and its energy and emotion. "It was the black music we dug," said Beatle John Lennon. They took that music and

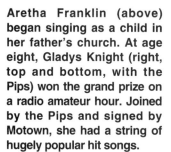

Aretha Franklin (above) began singing as a child in her father's church. At age eight, Gladys Knight (right, top and bottom, with the Pips) won the grand prize on a radio amateur hour. Joined by the Pips and signed by Motown, she had a string of hugely popular hit songs.

turned it into something that was all their own. "We didn't sound like the black musicians because we weren't black and because we were brought up on a different kind of music and atmosphere."

The Beatles were good; even people who didn't like rock were likely to agree about that. Even classical musicians were listening to them. It was hard for anyone to ignore the Beatles. But there were many other good musicians too. Aretha Franklin, the daughter of a Detroit Baptist minister, was one of the superstars. Franklin was another Hammond find. She managed to take gospel and blues and merge them together into something that was sad and raw and cool all at the same time—it was called *soul*. Franklin, like most of the rock stars, came from an ordinary background—it was her talent that was extraordinary.

Not everyone liked rock—in fact, some people hated it. The lyrics were often about sex or drugs, and the volume of the instruments—which were usually electrically amplified—could be earsplitting. The great jazz artist Duke Ellington said rock "had nothing to do with music." But a lot of people didn't agree, especially young people, who had those amplifiers turned way up.

Motor Town

Berry Gordy was a hard-working guy who drove performers mercilessly, but he had a great ear and he produced a fountain of black music that mainstream white listeners (and lots of others) ran to their record stores to buy. Gordy was from Detroit, which was then the auto capital of the world, known as Motown to many. So when it came to naming his record company—why, Motown it was.

Gordy started as a boxer, tried running a record shop, and worked on a Ford auto assembly line, but he hit the big time when he began producing records. He found talent in inner-city Detroit and turned it into gold. The Motown sound, with its pounding beat, dominated the radio waves. Gordy gave his discoveries makeovers and groomed them for pop stardom: Martha and the Vandellas, Smoky Robinson and the Miracles, the Temptations, the Supremes, the Jackson 5 (including Michael Jackson), Stevie Wonder, and Marvin Gaye were all Berry Gordy products. By 1977, his company, Tamla-Motown, was the largest black-owned conglomerate (a collection of smaller companies) in the U.S.

36 Nixon: Vietnam, China, and Watergate

The Nixons in Yorba Linda, California, where Richard (far right) was born in 1913. When he was nine, they moved to Whittier, where Richard's father ran a gas station and a grocery store.

During the Cold War years the presidents and the military kept many decisions from the American people. All kinds of things were called "top secret." Today we know that many of those "top secrets" were mistakes. In a democracy, the people should have the freedom to make their own mistakes—and learn from them. Democracy is based on an open government and an informed public.

Some years, like some people, stand out. Take 1492, or 1776, or 1860, or 1917, or 1945. You know what happened in each of those years, don't you? Well, find out if you don't, because each is noteworthy, significant, momentous, and consequential—which means they are not-to-be-forgotten years.

Now, 1968 doesn't rank up there with 1492—it wasn't *that* important. But, in the second half of the 20th century, it stood out as a pivotal year. And that means that things changed in 1968; they changed dramatically.

It was the year of those two awful assassinations. It was the year of the Tet offensive in Vietnam. *Tet* is the Vietnamese New Year. It is a big holiday—a kind of Christmas, New Year's, and Thanksgiving rolled into one. The North Vietnamese launched an attack during Tet; a lot of American soldiers were killed, and we realized we weren't winning that war (although our leaders had been telling us that we were).

Nineteen sixty-eight was also a year of urban riots and protests on college campuses. It was the year a computer named Hal starred in the movie *2001: A Space Odyssey*, and people gasped when they considered where technology might lead. It was an election year, and the end of a liberal era and the beginning of more conservative times. It was the year Republican Richard Milhous Nixon was elected president.

In 1962, Nixon lost the race for governor of California and told reporters, "You won't have Nixon to kick around anymore, gentlemen, because this is my last press conference." Six years later he ran for the presidency and won easily.

I have never thought much of the notion that the presidency makes a man presidential. What has given the American presidency its vitality is that each man remains distinctive. His abilities become more obvious, and his faults become more glaring. The presidency is not a finishing school. It is a magnifying glass.

—RICHARD NIXON, *MEMOIRS*

Richard Nixon played football in high school, but not in college. One observer said, "I've often thought with Nixon that if he'd made the football team, his life would have been different."

Words that *Richard Nixon never learned:*

When tempted to do anything in secret, ask yourself if you would do it in public; if you would not do it, be sure it is wrong.

—THOMAS JEFFERSON

Nixon's story begins way back in the 18th century, when a family named Milhous arrived in William Penn's colony. They'd come from Ireland and were Quakers: hardworking, peace-loving folk. Eventually the Milhouses moved to Indiana, where there were also many Quakers, and, after that, some of them went to California and helped found a Quaker town named Whittier (named after a 19th-century American poet. Who was—?). When the 20th century began, Frank Milhous was running a nursery (the kind where you raise plants, not children) in Whittier. He was prosperous and said to be a bit snooty. He wasn't at all impressed with Frank Nixon, the man his daughter Hannah chose to wed.

The Nixons, too, had come to America in the 18th century. They first settled in Delaware, about ten miles away from the Milhouses, across the border in Pennsylvania. But they probably didn't know each other.

Life was never easy for the Nixons. It certainly wasn't easy for Frank Nixon, who became an orphan when he was young and never got much love. People had mixed feelings about Frank Nixon: he could be thoughtful and kind, but he had a bad temper. Everyone agreed that Hannah was kindly; some called her a saint. Frank and Hannah had five sons and struggled to get by on Frank's modest earnings.

Their second son, Richard, a quiet, dark-eyed, serious boy, was the kind who never seemed to get his clothes dirty. Richard got good grades in school; in high school he learned to debate and to act. Then he went

Nixon went *to China in 1972. Here are his words describing what happened:* Chou En-lai stood at the foot of the ramp, hatless in the cold. Even a heavy overcoat did not hide the thinness of his frail body....I knew that Chou had been deeply insulted by Foster Dulles's refusal to shake hands with him at the Geneva Conference in 1954. When I reached the bottom step, therefore, I made a point of extending my hand as I walked toward him. When our hands met, one era ended and another began.

In 1971, the 26th Amendment to the Constitution was ratified. It said that 18-year-olds could vote. Are you getting ready? The most important job in the nation is that of citizen. This is a people's government. The people are in charge. The president, the governors, and the other public officials all work for us citizens. They are called *servants of the people*. But if you don't know enough about your government to understand what your servants are doing, they may rob you and steal your power.

This boy was a soldier in the Cambodian army. U.S. policy in Cambodia ended in disaster for the people, whether they were communists or not.

to Whittier College, was president of the student council, and got a scholarship to Duke University Law School (across the country in North Carolina). He was graduated in time to serve in the U.S. Navy during World War II.

It was politics that always seemed to interest him. So, as a young lawyer, when he got chances to run for Congress and then for the Senate, he grabbed them. Some people would never forget the campaign methods he used. They were unsavory—which means dirty. For one thing, he accused some of his opponents of being communists, when he knew they weren't. He ran against a woman senator and said she was pink right down to her underwear—because *red* and *pink* were words used to describe communists. That might have been amusing, but it was untrue. Someone on his campaign staff even forged a picture of the lady senator and a leading communist. Everyone was doing it, his supporters said of his mudslinging and dirty tricks. Do you think that excused him?

No question, Dick Nixon was bright and capable. In Congress he became known as a tough anti-communist, a kind of well-behaved Joe McCarthy. On most other issues, he sided with the moderates. As to civil rights, he was usually for them; in foreign affairs, he supported the Marshall Plan. He impressed people: he was smart, industrious, serious, and ambitious. Dwight D. Eisenhower asked him to be his vice president, and he did a good job in that office. People began talking about the two Nixons. One was very capable. The other Nixon didn't seem to care about truth and honor.

When he became president, he brought those two personalities with him. Richard Nixon the statesman talked of "law and order," and, after months of riots in our cities, that was just what most Americans wanted to hear. But the other Richard Nixon had no respect for the law when it affected him.

He claimed he had a plan to end the war, but he never said what that plan was. Then he kept us fighting in Vietnam for almost five more years (he was reelected in 1972). He took the war into neighboring Cambodia and Laos, without telling Congress that he planned to do it. He dropped more bombs than any president in our history, although he said he wanted to be a peacemaker.

The antiwar demonstrations had been bad when Lyndon Johnson was president; they were worse for President Nixon.

Nixon's intelligent, reasonable side helped him lead the nation in a new foreign-policy direction. Nixon was a pragmatist, which means a practical thinker, and he understood that the world was changing and that it was time to try to work with the communist nations. So he went to China (and took along three cargo planes, with 50 tons of television equipment, so the whole world could watch him walk China's Great Wall). He improved relations with that enormous nation. Then he went to Moscow (Russia's capital), the first American president to do so, and once again showed concern for world harmony.

Richard and Pat Nixon atop the Great Wall of China. Asked for his thoughts, the president said, "I think you would have to conclude that this is a great wall."

We got out of Vietnam much as we had gotten in—one step at a time. It was called *phased withdrawal*. But, after Saigon, the capital of South Vietnam, fell to northern forces, we finally withdrew completely. We had lost a war—although we didn't quite admit it. We were confused and humbled and weary. We needed to feel good about ourselves again, but something was going on at home that left us even more upset and dismayed. The problem, again, was one of leadership.

Something happened to Richard Nixon that is important for you to understand. It happened because of that distrustful side of his nature. He imagined enemies. He did what he wanted, and didn't worry about

Captain Denton Comes Home

Jeremiah Denton was the 13th American pilot to be shot down in the Vietnam War. That was back in 1965 (see page 129). Now it was seven years later, and a big C141 plane touched down at Gia Lam airport, outside Hanoi. The POWs were leaving the Hanoi Hilton, and Heartbreak, and Alcatraz. They were going home. (Some 50,000 Americans who had been in southeast Asia would not return home with them.) These men had endured great hardship and had survived. They didn't know it yet, but they were national heroes. They would help heal the wounds of war.

Jeremiah Denton

They flew east, to Clark Field in the Philippines. As the senior officer on the first plane of POWs to land, Jeremiah Denton—now Captain Denton—was asked to speak. The man who had suffered so much for so long stood straight and spoke clearly:

We are honored to have had the opportunity to serve our country under difficult circumstances. We are profoundly grateful to our commander in chief and to our nation for this day. God bless America!

They were words that made Americans proud.

WANTED

JAMES McCORD   DWIGHT CHAPIN   H. R. HALDEMAN   JOHN MITCHELL   JOHN ERLICHMAN

MAURICE STANS   EUGENIO MARTINEZ   G. GORDON LIDDY   CHARLES COLSON   HERBERT KALMBACH

JOHN DEAN   ROBERT MARDIAN   JEB MAGRUDER   RICHARD M. NIXON   BERNARD L. BARKER

VIRGILIO GONZALEZ   DONALD SEGRETTI   FRANK A. STURGIS   E. HOWARD HUNT JR.   HUGH SLOAN JR.

"This office is a sacred trust," said Nixon in 1973, the year he and dozens of others were investigated in connection with the Watergate break-in.

breaking the law, or about hurting people. He seemed to think that because he was president, he was above the law. But he was missing the whole point of American democracy. No one is above the law—not even the president. As North Carolina's Senator Sam Ervin, Jr., said, "divine right went out with the American Revolution."

Anyone who understands our democracy knows that the president is a servant of the people. Richard Nixon forgot that. He allowed his staff to play dirty, illegal tricks on his opponents. Burglars broke into Democratic Party headquarters and stole documents. Burglars broke into a psychiatrist's office and stole the confidential records of someone Nixon disliked. People tapped telephone lines and listened to private conversations. Money was gathered and used in illegal ways. Lies were told about people Nixon disliked. The government's tax office was used against his enemies. All of those things were against the law.

When some of that wrongdoing became known, people in the Nixon White House did something even worse. They paid hush money to keep some people quiet and to have others lie in sworn testimony to judges and juries. It was disgraceful. It was the bottom moment in the history of the presidency. It was called *Watergate* because the Democratic Party headquarters were in Washington's fancy Watergate apartments. Nixon's dirty-tricks workers burglarized those Watergate headquarters. They rented a room in a nearby hotel so they could spy on Watergate. (Americans spying on each other? The president involved! Sad but true.)

Shameful as it was, there was something positive about Watergate: our democratic system worked. When two reporters (Bob Woodward and Carl Bernstein) found out about the burglaries and the dirty tricks, they told of them. It took great courage to accuse a president and his

Tet was when American public opinion changed. After Tet, most Americans no longer supported the war.

aides. Their newspaper, the *Washington Post*, stood behind them. The press—the *fourth estate*—acted as it was meant to: as a responsible watchdog alerting the nation to danger.

Richard Nixon almost got away with criminal acts—but he didn't. The president was not above the law. Nor were other people in his administration. Vice President Spiro Agnew admitted to filing a "false and fraudulent" tax return. Agnew left office, was fined $10,000, and sentenced to three years' probation. Fifty-six men in the Nixon administration were convicted of Watergate-related crimes. Some went to jail. The Constitution writers had prepared

"I am not a crook," Nixon told reporters. His biographer Stephen Ambrose said of him, "Mr. Nixon wanted to become Richard the Great. He wanted nothing short of world peace and a prosperous, happy America. He was brought down by his own hubris, by his own actions, by his own character. This is tragedy." BELOW: By October 1973, Nixon's approval rating had hit a low of 17 percent. Ten months later, Republican senator Barry Goldwater informed him that he had lost almost all support in the Senate; Nixon resigned the next day.

The story continues on page 176.

Mercury, Gemini, and Saturn

Some 4 billion years ago, a small planet hurtled onto Earth and sent exploding pieces into the atmosphere. Those objects circled the earth, collided, collected, and became the moon. The earth and the moon eventually settled into a gravitational balance about 239,000 miles apart, with the moon orbiting Earth and its pull influencing the oceans' tides.

It was a long time before earthlings appeared; when they did, they watched the moon and the cycles of its appearance, and planted crops when the moon seemed to tell them to do it. They told stories of the moon, and dreamed by its bright, reflected light. So it was not surprising, when we actually pushed ourselves off the surface of the earth, that the moon was where we wanted to go.

It was an outrageous idea, to expect to leave the earth's atmosphere and make it to that distant globe, especially in the very century that people had first learned to fly.

We might not have tried it at all if it hadn't been for Russia. When the Russians sent a vehicle into space—called *Sputnik*—we couldn't quite believe it. We Americans had the idea that we were better than others. It was a kind of national arrogance. We aren't better or smarter than other people. (What we have is a terrific idea—for free government—that is the envy of a lot of other nations and has helped most of us pursue happiness.)

But scientific achievement? We have to work as hard as anyone else to make and do things. Russia's *Sputnik* got us energized. We didn't want our communist foes to take over space.

Then, in April 1961, the Russians sent a man rocketing into space. His name was Yuri Gagarin (guh-GAR-in), and he had a boyish grin and a lot of courage. When he came back to earth he landed in a field where he startled a cow and two farm workers. "Have you come from outer space?" stammered Anya Takhtarova to the man in the orange flight suit. "Yes. Would you believe it, I certainly have," said cosmonaut Gagarin.

The United States had a space agency, NASA (the National Aeronautics and Space Administration), and a space program—but we were behind the Russians, and we couldn't stand that idea.

President John F. Kennedy made a speech announcing our intention to put a man on the moon "before the decade is out." We were off on a space race.

What would life be like in space? On the earth, it is the pull of gravity that keeps your legs on the ground. But when there is zero gravity—as in space—there is no pull. You float around. You have no weight. If you eat a cookie in space, the crumbs float. If you want to sleep in a bed, you have to be strapped down. Other things have to be considered. Normal breathing is impossible in the vacuum of

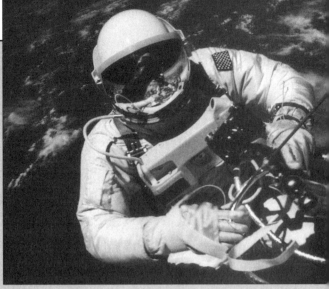

Edward H. White II floats in space in a specially designed spacesuit and helmet, which has a gold-plated visor to protect him against the sun's rays. In his hand is an HHSMU (a "hand-held self-maneuvering unit") with which he controls his movements in space.

space; a spacecraft or spacesuit has to be equipped with its own atmosphere.

A trip to the moon would be a voyage like the one that Columbus made. No one knew where it might lead. Would we create colonies in space? Would we mine the moon's resources? Would we put factories in space and return the earth to its gardenlike heritage? Would we explore other galaxies? Would we meet other beings out there?

This moon trip became the will of a nation. It took the talent of thousands of brains, it took the lives of some astronauts (who were killed in explosive misfires), and it cost $25.5 billion, which came from the earnings of America's citizens.

The first step toward the moon was a flight into space. Alan B. Shepard was squeezed into a spacesuit in a space capsule just big enough to hold him. This was the Mercury project, named for the swift messenger of the ancient Roman gods. (The craft was called *Freedom 7*).

Some smart scientific brains worked on Project Mercury, but they forgot something important: astronauts have to go to the bathroom. Shepard had a big breakfast the day of his flight. Finally, he had to go in his spacesuit. Then he was off, blasted into space with a great roar and arched back to his home on earth with a mighty splash into the ocean. The flight lasted 15 minutes. (Today's space shuttles have toilets. Flowing air substitutes for gravity and draws wastes into storage containers. Astronauts now wear everyday clothing inside the space vehicle. Spacesuits are needed only for activities outside the spaceship.)

The Mercury flights—there were six of them—were an important first step. One Mercury flight lasted 34 hours.

Next came *Gemini*, named for twin stars. They were two-man flights intended to test rendezvous (meeting) and docking techniques. Gemini met a target vehicle, named *Agena;* the spacecraft touched noses and clamped themselves together. The Gemini astronauts walked outside the capsule—into outer space—but with a cord that firmly tied them to their vehicle.

The Gemini spacecraft was a big improvement over Mercury. It was bigger and could be steered by the astronauts.

On the morning of July 16, 1969, five months before President Kennedy's deadline of the end of the decade (Richard Nixon was now president), the sun was bright and the skies were clear at Cape Canaveral on Florida's east coast. Some 8,000 people were packed into a special viewing area; others jammed nearby roads and beaches. Photographers in TV helicopters flew overhead taking pictures of the crowds and of the good-luck messages written in beach sand.

Nearby, three men sat strapped elbow to elbow inside a narrow capsule on top of a rocket that stood as tall as a 30-story building. Neil Armstrong, a civilian pilot, was in the left seat. Some said he was the nation's best jet test pilot. Armstrong had the personality of a cowboy-movie hero: cool. Edwin E. Aldrin sat in the middle. Everyone called him by his school nickname, Buzz. Buzz Aldrin was an air force colonel with a big brain. Some of his scientific ideas had gone into this mission. Michael Collins, another air force officer and test pilot, was to pilot the command ship, which would orbit the moon while the other two men descended to the lunar surface in the landing vehicle.

The rocket—named *Saturn*—belched fire and its own billowing clouds, lifted off, and seemed to rise slowly. But that was an illusion; after two and a half minutes *Saturn* was 41 miles above Earth. It was traveling at 5,400 mph (miles per hour) when its first stage fell away. (How fast can an automobile go? How about a commercial jet?)

The next stage took the astronauts 110 miles above Earth, carrying them at 14,000 mph, and was jettisoned (dropped away). The third stage got them to 17,400 mph; they were now weightless and orbiting the earth. It was 17 minutes after liftoff. After they

had circled the globe twice, the third-stage engine fired the ship away from Earth's orbit. "It was beautiful," said Armstrong. He was cruising toward the moon. It would take three days to get there.

(Three days was the time it took Thomas Jefferson to make the 90-mile trip, in a horse-drawn carriage, from his plantation at Monticello to his plantation at Poplar Forest.)

Everyone on Earth went on this trip. Television took us into space and then put us on the rocky, craggy, pock-marked moon. When two men stepped out of the landing vehicle, we were there—all the peoples of the earth. It was an American spaceship, but it was a world event.

Neil Armstrong stepped onto the moon's crunchy soil and said, "One small step for man, one giant leap for mankind." It was an understatement. The man in the moon was now real, and we were standing with him.

The view from the moon was of one Earth—it was not one of small, separate nations. Perhaps the next bold journey would be one that the united nations of the earth would take together.

Advice from the Past

A president is impeachable if he attempts to subvert the Constitution.
—JAMES MADISON

Education is the true corrective of abuses of constitutional power.
—THOMAS JEFFERSON

The fate of empires depends on the education of youth.
—ARISTOTLE

The boys of the rising generation are to be the men of the next, and the sole guardian of the principles we deliver over to them.
—THOMAS JEFFERSON

It is while we are young that the habit of industry is formed. If not then, it never is afterwards.
—THOMAS JEFFERSON

for this kind of emergency by giving Congress the power to impeach and try a president. (See Book 7 of *A History of US* to read of the impeachment of Andrew Johnson.)

In the House of Representatives, articles of impeachment were prepared. President Nixon was charged with lying, obstructing justice, and using the Internal Revenue Service (the tax office) and other government agencies illegally. Nixon was going to be impeached. After that, he would face a trial in the Senate for "high crimes and misdemeanors." He chose to leave the presidency instead. He resigned as president of the United States (the only man ever to do so).

In England, an editor of the London *Spectator* wrote that the U.S. presidency had gone from George Washington, who could not tell a lie, to Richard Nixon, who could not tell the truth.

Barbara Jordan Examines the Constitution

Representative Jordan, a lawyer, and the first black woman elected to the Texas State Senate, was a member of the House Judiciary Committee that considered the impeachment of President Nixon for "high crimes and misdemeanors" (see Article II, Section 4, and Article I, Sections 2 and 3, of the U.S. Constitution). Here is an excerpt from her impassioned speech, which was carried on national television and earned her the role of keynote speaker at the 1976 Democratic National Convention.

Mr. Chairman...Earlier today we heard the beginning of the Preamble to the Constitution of the United States, "We, the people." It is a very eloquent beginning. But when the document was completed, on the 17th of September in 1787, I was not included in that "We, the people." I felt somehow for many years that George Washington and Alexander Hamilton just left me out by mistake. But through the process of amendment, interpretation, and court decision I have finally been included in "We, the people."

Today, I am an inquisitor....My faith in the Constitution is whole, it is complete, it is total. I am not going to sit here and be an idle spectator to the diminution, the subversion, the destruction of the Constitution.

"Who can so properly be the inquisitors for the na-

tion as the representatives of the nation themselves?" (*The Federalist*, no. 65). The subject of its jurisdiction are those offenses which proceed from the misconduct of public men. That is what we are talking about....It is wrong...to assert that for a member to vote for an article of impeachment means that that member must be convinced that the president should be removed from office. The Constitution doesn't say that....In establishing the division between the two branches of the legislature, the House and the Senate, assigning to the one the right to accuse and to the other the right to judge, the framers of this Constitution were very astute. They did not make the accusers and the judges the same person.

"If you're going to play the game [of politics] properly," said Barbara Jordan, "you'd better know every rule."

What is Barbara Jordan's point? Can you put her thoughts in your own words? What do these words mean: **inquisitor, diminution, subversion, astute?**

37 A Congressman and a Peanut Farmer

"I am a Ford, not a Lincoln," said Gerald Ford when he was sworn in as vice president. He was a keen golfer and did push-ups and swam daily in the White House.

Gerald Ford was never elected president or vice president, and yet he became president of the United States. How did that happen?

This was the way: President Nixon chose him to replace Spiro Agnew when Agnew resigned as vice president. Then, when Nixon resigned, Ford became president.

Ford, a popular, pleasant man, was a congressman from Michigan and House minority leader. That means he was the Republican leader in the House of Representatives (where there was a Democratic majority). When he became president, he put his feet into two hornet's nests: the messes that were left from Watergate and Vietnam.

He said he would heal the "long national nightmare," and he was talking about the scandal of Watergate. He promptly granted Nixon an unconditional pardon for any wrongdoings against the United States. Some people howled in protest—they thought Nixon should be put on trial—but others believed the country was better off spared that agony; they were glad to forget the national nightmare. Then Ford pardoned draft protestors who had refused to fight in Vietnam, although there were some conditions attached to their pardons.

During Gerald Ford's presidency the last U.S. troops and support workers were evacuated from Vietnam. Vietnam would

"To me," said Ford, seen here waving goodbye to Nixon as he left the White House for good, "the presidency and the vice presidency were not prizes to be won but a duty to be done."

177

> I believe that truth is the glue that holds government together, not only our government but civilization itself....As we bind up the internal wounds of Watergate, more painful and more poisonous than those of foreign wars, let us restore the golden rule to our political process, and let brotherly love purge our hearts of suspicion and of hate. —GERALD FORD,
> IN HIS INAUGURAL ADDRESS

take its place in history as America's worst foreign-policy defeat.

As president, Gerald Ford didn't break new ground or excite the imagination of most Americans. His wife, Betty Ford, did. She spoke out openly on controversial subjects, especially the rights of women. She talked of her own personal problems with cancer and alcoholism and discussed the pressures on young people to use drugs. But that wasn't enough to get her husband elected. In 1976, when Ford tried to win an election for president, he lost.

Betty Ford aged 20

James Earl "Jimmy" Carter became the 39th president. A redheaded peanut farmer with a big, toothy grin, Carter had graduated from the U.S. Naval Academy at Annapolis and become governor of Georgia. When he decided to run for president he was hardly known outside his state. Most people laughed at the very idea; his own mother laughed. But Jimmy Carter was determined. Soft-spoken and deeply religious, Carter told the American people, "I will not lie to you." And, as far as we know, he never did.

Carter was a southern Democrat with progressive views on civil rights and moderate ideas on economics. But he was an outsider when it came to dealing with the government in Washington. He brought his friends from Georgia with him to the capital. They had some good ideas, and President Carter thought Congress would go along with those ideas. But Jimmy Carter hadn't learned the ways of Congress. He couldn't get things done. It was frustrating for him and for the country.

Besides, he was unlucky. While he was president, a worldwide energy crisis made prices—especially the price of oil and gas—in the United States zoom way

The Carter family—from left, Amy, Jimmy, and Rosalynn—at home in Plains, Georgia.

No president's child had gone to public school since the days of Theodore Roosevelt. But Amy Carter did. She walked to public school with a bookbag on her back. (The Secret Service followed her.) Amy roller-skated on the porch in front of the White House and played in a treehouse her father had built for her.

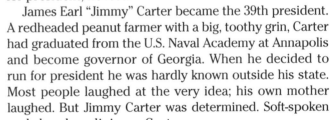
In 1979, with 6 percent of the world's population, the United States used 33 percent of the world's energy. Gas prices soared, and miserable commuters waited hours at pumps.

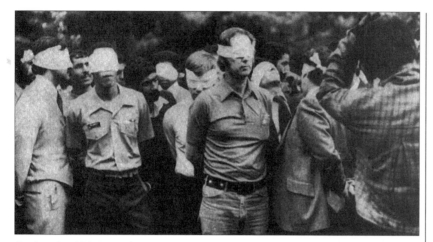

During the 444 days of the American hostages' imprisonment in Iran, their captors broadcast scenes such as this—where the hostages were blindfolded—to American TV networks. It was the single worst disaster of Carter's presidency.

In 1978 President Carter posthumously restored American citizenship to Jefferson Davis. (What does *posthumously* mean? Who was Jefferson Davis, and why was his citizenship taken from him?)

up. (It was an inflationary time.) Then the ruler of Iran, the shah, was overthrown and replaced by a fundamentalist Muslim religious leader, the Ayatollah Khomeini, who preached hatred of the United States. The Iranians captured some Americans and held them as hostages for 444 days. President Carter ordered a daring rescue mission, but it fizzled into an embarrassing mess of poor planning and failed equipment. As if that weren't bad enough, the Soviet Union invaded Afghanistan, another Muslim country in Central Asia, and, when we protested, relations with Soviet Russia became icy.

A Muslim (or Moslem) is a believer in Islam, a religion that sees Mohammed as the chief prophet of God.

President Carter did serve as a peacemaker between Egypt and Israel. And he did get Congress to agree to turn the Panama Canal over to Panama at the end of the century (which was a new Good Neighbor policy). And he did support measures to help protect our natural environment.

But Jimmy Carter never seemed to capture the enthusiasm of most Americans. A highly intelligent, compassionate man, he was, nevertheless, a poor communicator. He tried to solve problems of national debt and energy conservation by asking people to make sacrifices. Maybe he didn't know how to ask—or maybe Americans weren't ready to make sacrifices. When Carter ran for reelection, he was defeated.

In 1978, Carter invited Egypt's president, Anwar Sadat (left), and the prime minister of Israel, Menachem Begin (right), to a summit meeting at Camp David, the presidential retreat in Maryland. Carter persuaded the warring leaders to sign an accord respecting one another's borders and agreeing to a peace treaty.

179

38 Taking a Leading Role

Ronald Reagan in his pre-presidential years and one of his less distinguished movies, *Bedtime for Bonzo*; his co-star was a chimpanzee.

Ronald Reagan was a dream come true for the leaders of corporate America—a politician who is a real actor; a man who started his career by convincing radio audiences that he was actually at a sports event when he was really reading off a ticker-tape machine and inventing the details; a man who made his living following directions and reading from scripts and convincing the American people that the resultant performance was reality. Is it really surprising that Ronald Reagan should graduate from fake broadcasts (which his listeners enjoyed) to fake statistics and fake facts?

—DAVID WALLECHINSKY AND IRVING WALLACE,
THE PEOPLE'S ALMANAC #3

The next president was a great communicator—in fact, that was what people started calling him. He had been a movie star and a television personality, and he knew how to use the media as no president had before. His name was Ronald Reagan, and he was in his seventies during most of his presidential years. Although he was old enough to be your great-grandfather, few people thought of Ronald Reagan as an old man. He seemed boyish, easygoing, likable, and friendly.

After the turmoil of the '60s and '70s, and the boring presidencies of Ford and Carter, this television president seemed reassuring. He was cheerful, optimistic, and—although he campaigned on a platform of thrift—he spent a lot of government money. The '80s, when he presided, were a high-living time. The rich got very rich, and the poor got much poorer.

Ronald Reagan set the tone for that decade right when it began—with his inauguration on January 20, 1981. It was the fanciest, most expensive inauguration in American history, costing five times more than the Carter inaugural. It was a black-tie, mink, and diamond affair, with limousines and designer dresses everywhere.

Andrew Jackson had invited the common people to his inaugural and they had poured into the White House. At Jimmy Carter's inaugural, almost all the events were open to the public, and most of them were free. Reagan's inauguration was by invitation only, and invitations were expensive. One inauguration planner said it was all about "class and dignity." Harry Truman would have been horrified, but many Americans seemed awed by the glitter and glamour.

Calvin Coolidge's portrait was brought down from the White House attic, dusted off, and hung in the Cabinet Room, right near the president's chair. (Portraits of Truman and Jefferson came down from the wall.) Coolidge, a Reagan hero, had said, "The chief business of the American people is business." The new chief executive agreed. This president was about to make some radical changes in the nation's economy, but he did it in such a pleasant manner that at first hardly anyone noticed.

Reagan had strong convictions, and he got things done. He had campaigned against taxes, against "big government," against government regulations, and against communism. Those all seemed like good things to be against.

Almost as soon as Reagan became president, he changed the nation's tax

Nancy and Ronald Reagan at one of their inauguration parties. "Thank goodness it's back," said the *New York Times*, "that froth in the confection of language, that lovely whipped cream of a word—luxury."

structure. He drastically reduced taxes on the rich, explaining that this would leave the rich with more money to invest in businesses, and that would help everyone. It was Coolidge's *trickle-down economics*, now called *supply-side economics*. George Bush called it *voodoo economics*. (That was before Bush became Reagan's vice president.) During the Reagan years, much of the nation's tax burden was transferred from the wealthy to the middle class. The income of the poorest fifth of the nation's families went down 6.1 percent during the Reagan presidency; the income of the highest-paid Americans went up 11.1 percent.

Supply-side economics didn't work. Reagan was spending more money on arms and weapons than had ever been spent before, and the government's income from taxes wasn't keeping up. When you spend more than you earn, you have to borrow money and you go into debt. You have a

Ronald Reagan was born 11 years after the 20th century began, and his two-term presidency ended 11 years before the beginning of the 21st century.

During the Reagan years, poverty and homelessness often went together. Some of the homeless had lived in hospitals for the mentally ill, many of which now closed their doors to people not considered a danger to society. Unable to cope alone, some ended up on the street.

Assassination Attempt

Shots were fired, and a Secret Service man pushed President Reagan to the floor of the limousine. "Jerry, get off me. You're hurting my ribs," he said. But a bullet had gone through his lung and was three inches from his heart. That was why his ribs hurt. When he coughed, blood came up, and the limousine sped for the hospital. As the president was being wheeled to an operating room, he saw his wife, Nancy. "Honey, I forgot to duck," he said. When he saw the doctors who were removing the bullet he said, "I hope you fellas are Republicans." Ronald Reagan had spunk (and a sense of humor).

deficit (DEF-uh-sit). The nation already had a big deficit; during the Reagan years the deficit turned into an out-of-control monster. The deficit became GIGANTIC. It rose from $58 billion in 1981 (when Reagan became president), to $220 billion in 1986, to $2.3 *trillion* in 1988 (when Reagan's presidency ended). When you owe money you not only have to pay it back, you have to pay interest on the loan. When you leave a debt, you are asking future generations to pay your bills. You and your children and grandchildren will have to pay for the Reagan deficit. You'll pay for it in your tax bills.

Reagan feared and hated communism and called Russia an "evil empire." His hatred of communism made him determined to build up our military forces. In the first five years of his administration, defense spending—military spending—rose about 400 percent. That not only added to the debt, it also widened the gap between some of the rich and poor regions in the country. Everyone paid taxes, but some regions got large military contracts (which gave them business and prosperity), while other regions received little government money and declined.

President Reagan said he hated big government. "Government is not the solution to our problem," he said. "Government is the problem." He cut back on many government programs, especially Lyndon Johnson's Great Society anti-poverty programs. Money for housing, food stamps, school lunches, and environmental protection was reduced drastically. Education funds were cut by almost 50 percent from Carter's proposed budget. With the federal government doing less, city and state governments had to do more. They grew enormously during the Reagan years. The executive branch, which the president controlled, grew too.

Reagan did away with much government regulation, especially in banking and television. That means there was little public control of those fields. Without regulators watching out for the people's interests, scandals and corruption in the banking industries ballooned. Many ordinary investors—people who put their savings into banks that misused and wasted the money—suffered. An enormous amount of tax money was needed to bail out those industries. You and your children and grandchildren will pay for this, too.

Does the government have the right to regulate TV? Many Americans seem to think so, but others don't. During the Reagan years, limitations on the number of commercials that could be aired during a certain amount of viewing time were lifted. Commercials soon filled TV screens.

In his budgets, Reagan embraced defense spending and cut taxes for businesses and people with high incomes. He reduced federal money for welfare programs and aid to big cities.

In the '80s, entertainment stars, sports figures, television announcers, and big spenders became the nation's chief celebrities. People who made a lot of money—however they did it—were held in awe; wheeler-dealers who manipulated financial markets were admired. One financial big shot, speaking at Stanford University's business school, said, "Greed is good." (He ended up in jail.) Our inventive geniuses—people like Ben Franklin, Cyrus McCormick, Thomas Edison, and Henry Ford—had made American prosperous by making good products. But the moneymakers of the '80s didn't make products; they just made money.

In 1987, Mike Milken, of the Wall Street investment firm Drexel Burnham, made $550 million in bonuses for selling "junk bonds."

It had all happened before in the United States. A hundred years earlier, in the 19th century's Gilded Age (the 1880s and '90s), people worshiped the dollar, and money barons were idolized. It happened again in the 20th century's Roaring Twenties, when getting and spending money seemed more important than making people's lives better.

George Washington, James Madison, and Thomas Jefferson had seen government as a noble calling. John F. Kennedy made outstanding people eager to work in Washington. But both Presidents Carter and Reagan talked about government

Little Wars

The Reagan administration got involved in civil wars in Lebanon and Nicaragua, bombed Libya, and invaded the small island of Grenada.

The CIA backed the Nicaraguan contras "not only against the government in place, but against the expressed will of all the neighboring states except our own questionable clients, Honduras and El Salvador."

"Almost everything went wrong, in the invasion of Grenada, that could have. The war was won because it could not be lost—the American invaders had a ten-to-one superiority over the defenders, and all of the air and artillery weapons used in the 'war.' Two-thirds of American casualties were inflicted by other Americans or by accident. The exclusion of reporters from the scene helped suppress knowledge of these facts."

—GARRY WILLS, FROM *REAGAN'S AMERICA AND INNOCENTS AT HOME*

In 1985, Mikhail Gorbachev (left) took power in the Soviet Union and began to relax its totalitarian regime. In 1987, he and Reagan signed the first arms-control treaty to agree to destroy nuclear weapons.

How are you going to remember all the presidents after World War II? Here are their names: Truman, Eisenhower, Kennedy, Johnson, Nixon, Ford, Carter, Reagan, Bush, and Clinton. Just write out their first initials and use them to make a phrase or sentence. Here is an example: *The Evil Kryptonite Just Nailed Foolish Clark—Real Big Catastrophe!* Maybe you can write a better sentence.

as if it were a bad thing, and that made some people reluctant to work for the government. Which was harmful, because, like it or not, we need big government. *We are a big nation.* But we need big government that is efficient and considerate of our needs. In the second half of the 20th century, our government became costly and inefficient, and it didn't consider the needs of many ordinary Americans.

During the Reagan years, some American officials secretly and illegally sold weapons to the terrorist government of Iran. Secrecy doesn't belong in a democracy. Lying doesn't, either. The whole point of democracy is that it is government by the people. How can the people control their government if that government deceives them?

Before Reagan left office, 138 members of his administration had been officially investigated for criminal misconduct. Many were convicted. Scandals at the Department of Housing and Urban Affairs cost the taxpayers billions of dollars. Conservative columnist James J. Kilpatrick wrote, "For the eight years of [Reagan's] administration...the president paid virtually no attention to this huge, costly department." Haynes Johnson, who won a Pulitzer prize for his reporting, said, "The record of his administration was the worst ever."

But Ronald Reagan's self-confidence, and his humor and gusto, made people feel good. He was a very popular president. Warren Harding, who was president during the frantic and fun 1920s, had also been very popular. His administration, too, was full of scandal and mismanagement. Neither Reagan nor Harding was ever said to be personally involved with those scandals, but the country's citizens paid a terrible price for their mismanagement.

In Reagan's second term, the cracks in the economy of the Soviet Union began to split apart. This once-mighty nation was on the edge of collapse. The United States began to relax. Reagan was a flexible negotiator. Relations between the two powers began to get friendly.

In 1989, Reagan's vice president, George Bush, became the nation's 41st president. The son of a Connecticut senator, Bush was the youngest pilot in the navy during World War II. After the war he moved west, to Texas, and became an oil man. But it was government that fascinated him, and he served in a series of important jobs—right up to the vice presidency.

When he was elected president, he inherited big problems. America's cities were decaying; our schools were behind those in many other nations; crime was epidemic; and the debts of the high-living '80s landed on his desk. Fifty years of war preparations had taken their toll on us, too. The American economy turned sour. People lost their jobs; there was a recession. What was happening? Almost everyone was worried.

39 Living on the Edge

In the 1940s, Tyson's Corner, Virginia, was just that: a corner where two country roads met. Today, it is a concrete, steel, and glass edge city.

Do you remember right after the war (at the end of the '40s), when Bill Levitt built Levittown, with its look-alike houses?

It was a long drive from that potato-field suburb to the big city where most jobs were located. At first you didn't mind, because the highways weren't too crowded, but as more and more suburbs got built, the roads became turtle slow. The solution was to build more roads. But every time a new highway was built, there would be new suburbs—with everyone crowded onto those highways, heading for the city and the things the city held: jobs, shopping, and entertainment.

Then something began to happen. It was the well-off executives who lived in fancy suburbs near the big cities who seem to have started it. They didn't like that commute now that there were added cars on the road. So they moved their work out to where they lived—usually to pretty, green areas on the edge of the city.

All of this started happening at a good time to make changes. We had been a nation of industrial workers. But factories

Technology has always changed the way people work. Now it is changing where they work, too.

> **More than anyplace else, California became the symbol of the postwar suburban culture. It pioneered the booms in sports cars, foreign cars, vans, and motor homes, and by 1984 its 26 million citizens owned almost 19 million motor vehicles and had access to the world's most extensive freeway system.** —KENNETH T. JACKSON, *CRABGRASS FRONTIER*

Zones for Clones

Alex Marshall, who writes on urban affairs, says:

As the suburbs become the norm, all the problems of the norm are becoming part of the suburbs. There have been some studies done that show income dropping and crime rising faster than in center cities.

Exclusionary zoning is a bedrock of suburbia and edge cities and people often overlook it. The suburbs have systematically kept the poor and lower working classes out of their cities by refusing to build high-density housing. You have to get to a certain income to afford suburbia. It's designed that way, to keep the "riffraff" out. Virginia Beach has few racial problems. Not because it doesn't allow black people in, but because it allows only middle-class black people in. That's the main reason its suburbs (a lot of them) have integrated so easily.

Zoning laws say what kinds of buildings can be built and where they can be built. (You can't turn your house into a restaurant if its zoning says your block is residential.)

Cities are always created around whatever the state-of-the-art transportation device is at the time. If the state of the art is sandal leather and donkeys, you get Jerusalem....When the state of the art is carriages and oceangoing sail, you get the compact, water-dominated East Coast cities of Paul Revere's Boston and George Washington's Alexandria....Canal barge and steamship give you Boss Tweed's New York....[Railroads yield] Chicago. The automobile results in...Los Angeles.

When, in 1958, you threw in the jet passenger plane, you got more Los Angeles in strange places—Atlanta, Denver, Houston, Dallas, and Phoenix.

The combination of the present is the automobile, the jet plane, and the computer. The result is Edge City.

—JOEL GARREAU, *EDGE CITY*

were getting fewer, and the new work was likely to be thinking work.

In 1952, more than half of our workers held manufacturing jobs; by 1992, the figure was less than 18 percent. We were on our way from the Industrial Age (with its factories) to the Information Age (with offices).

We were still a great industrial nation—we just didn't need as many people to make things. Our steel industry had become so efficient it could make a ton of steel with fewer man-hours than in any other nation. Instead of getting a job on a production line, you were likely to find work as a computer specialist or as an engineer. People without a good education were having a hard time getting good jobs. Many of them were stuck in the old cities—and out of work.

Meanwhile, as the business leaders moved their offices and office workers to new headquarters and office parks on the city's edge, other jobs appeared in those same locations. After all, the office workers had to be fed, clothed, and have their children cared for. So lots of Americans began moving—to the edge, to new *edge cities*—where jobs, shopping, homes, and entertainment were all appearing.

By the mid-1980s there was more office space on the fringes of New York than in the big city itself. The same thing was happening across the country. Edge cities (some called them *superburbias*) were a new phenomenon, and growing like dandelions after a rain.

How do you recognize an edge city? Joel Garreau, who wrote a book about them, says they are places where the population *increases* at 9 A.M. on workdays. Edge cities have a whole lot of office space (at least as much as downtown Seattle has); they are near a traditional city, and near an airport too. They are almost always near an intersection of highways; they usually hold several giant shopping malls, and a variety of housing. But most edge cities seem to be designed more for cars than for people. The critics say that few of them have any kind of a soul—or the feel of community.

Do you live in an edge city? You can find them all

across the nation. Atlanta has edge cities at Midtown, Buck-head, Perimeter Center, and the Cumberland Mall–Galleria area. If you live near Phoenix, consider Scottsdale, Camel-back–Biltmore, and Uptown–Central Avenue. Houston has the Galleria area, Westheimer–Westchase, Greenspoint–North Loop, the Rice University–Texas Medical Center area —and more.

New Jersey is filled with edge cities; they have larger populations than the old cities of Newark and Trenton. One New Jersey edge city, around the intersection of interstate highways 78 and 287, is headquarters to some big compa-nies that make products—from telephones to toothpaste—that you use every day. Those headquarters are mighty swanky, with gyms, tennis courts, hanging gardens, and ele-gant dining rooms.

Our traditional cities with their downtowns—like Chicago, Kansas City, Duluth, Dallas, or Milwaukee—were all built before the 20th century began. Those big cities are granddaddies—and they are not reproducing. The edge cities are kids; most were born in the last quarter of the 20th century.

Are they the way of the future? Don't bet on it. Things are happening fast. Information-age technology keeps changing the way we work and live. Today, lots of people are taking their jobs home. Many are moving to rural areas, plugging in a fax, a computer, and a modem—and going to work. That trend is expected to continue.

But the edge cities aren't standing still. The newest ones are being planned with grassy malls, meeting rooms, gazebos for outdoor performances, and streets for walking—all intend-ed to foster public activities and a sense of community.

And those old cities? Well, they're changing, too. In Denver, inner-city warehouses are being turned into apartment lofts. The city's core is vibrating with re-vamped shops, trolley cars, a sunny outdoor mall, and a big in-town amusement park. People are choosing to come back downtown to live.

Urban designer Todd Johnson says, "Balance is the key. A healthy city grows in all areas—it offers a wealth of choice—that's what makes great urban areas great. Seeking your own niche in the supermarket of choice is the fun of it all."

Beneficial Management, which sells in-surance, built its Peapack, New Jersey, headquarters—at the intersection of highways 78 and 287—in a style remi-niscent of an Italian village (an Italian vil-lage with tennis courts, a gymnasium, and underground passages).

According to Joel Garreau, "Eighty-eight percent of Americans live outside what has traditionally been defined as a big city....a place ...with half a million population."

Malling: hanging out on weekends, 1990s-style.

40 The End of the Cold War

In 1989, the Soviet Union—and its monuments, such as this statue of the founder of the Soviet secret police—finally came crashing down.

In 1994, two old men met in Vietnam. Admiral Elmo Zumwalt, Jr., who had commanded U.S. naval forces in southeast Asia, faced Vietnam's General Vo Nguyen Giap. Each was a legendary leader. Each had lost loved ones in that cruel war. They had been bitter enemies. Now, for the first time, the old warrior-foes looked into each other's eyes. They shook hands; then they embraced. Giap autographed his book *People's War, People's Army*, and gave a copy to the admiral. Zumwalt signed *On Watch*, his tale of his wartime experiences, "With respect to a former adversary and friend."

Something had happened to America in the turbulent years of Vietnam and civil rights activism.

"Somewhere, in the decades of upheaval, came a wrong turning," wrote Theodore White, a historian of the presidency. Many Americans felt the same way. Back when Eisenhower and Kennedy had run for the presidency, most people trusted the national government. Local politicians were sometimes called crooks (and some were), but the president was above all that. You might not like his politics, but he was the president, and someone to admire, or at least respect.

The president and Congress seemed to have shrunk. Everyone agreed: there was a need for inspiring leadership, especially with all the astonishing changes in the world. Satellite communication, computers, fax machines, and jet travel had made the world shrink. Clearly, all the earth's people were passengers on the same global spaceship. Ideas born in one country could be used minutes later in another. When one country had a money crisis, pocketbooks hurt elsewhere. We were all becoming linked.

The European nations had joined together to form the European Community (EC) or Common Market. That means their businesses acted as if Europe was all one nation. That made sense; it gave them collective power. Japan was making excellent products—cars, computers, TV sets—and had become a leader in the world business market. Nations like Korea and China were making products that people wanted, and they were also producing capitalists. When World War II ended in 1945, the United States didn't have much competition—we were the supplier of goods and farm produce to much of the world. That had changed.

All of this demanded new thinking about business and economics, but it didn't change our founding ideas of democracy and freedom. They seemed better than ever. People all over the world wanted our kind of democratic government. When students in China rebelled against their dictatorial government, they marched around with a model of our Statue of Liberty. The Chinese dictators squashed that uprising; the Chinese students didn't get what they wanted—freedom—but they all knew that a government can't continue forever without the support of its people.

In the 20th century we'd fought hot wars and a cold war—all for our democratic way of life. It had triumphed. That became clear to everyone in 1989, when the Soviet Union broke into pieces.

Yes, the U.S.S.R., the Union of Soviet Socialist Republics, the land we called Russia—a nation composed of many states—just fell apart. Communism, as a political system, was a failure. It had begun with high hopes as a visionary experiment. The experiment hadn't worked. Anyone who dared speak out against the government was sent to a *gulag* (prison camp). Marx's economic ideas hadn't worked either. Government ownership of land and its products had wrecked Russia's economy. People work hard to get a reward—usually more pay. But under communism you weren't paid for your output. So people did as little as possible. Fewer things were produced, and the Soviet Union grew poorer and poorer. Like us, Russia was spending vast amounts of money on weapons; its crippled economy couldn't take the strain.

Finally, the Russian people just threw communism out. It was stunning, and it was peaceful. It meant that everything had changed in the world's politics. *The Cold War was over.*

Americans had spent all

> As the free world grows stronger, more united, more attractive to men on both sides of the Iron Curtain—and as the Soviet hopes for easy expansion are blocked—then there will have to come a time of change in the Soviet world....Whether the communist rulers shift their policies of their own free will—or whether the change comes about in some other way—I have not a doubt in the world that a change will occur. I have a deep and abiding faith in the destiny of free men. With peace and courage, we shall someday move on into a new era.
>
> —GEORGE BUSH

At the end of the 1980s, the United States seemed to be buying much more from the Japanese than we were selling to them; one American success came in 1992, when President Bush opened a branch of the U.S. company Toys R Us in Japan.

In 1989, thousands of Chinese students demonstrated in Tiananmen Square in the center of Beijing, the capital of China, demanding freedom to think and speak and live. To symbolize that freedom they made their own Statue of Liberty. The government cracked down, broke up the demonstrations, and persecuted and killed many of the rebels.

There are some 1,290 different practicing religions in the United States. In this country people of all faiths live together in relative harmony. *We have never fought a war over religion.* Most people agree that is because of our First Amendment—it separates church and state, which means that our government keeps its hands off when it comes to religion. That isn't true in most other countries. At the beginning of the 1990s, more than 30 religious wars were being fought around the globe.

the years since World War II battling communism. Now there was no giant to battle. Sometimes we acted as if we didn't know what to do.

It was President George Bush who began taking us in a new direction. When the dictator of Iraq sent troops into neighboring Kuwait and took over that nation, Bush led a forceful response. The United States, with the United Nations, stopped that aggression with a powerful, short war.

The next time President Bush called out American forces, it was to help starving people. Somalia, which elbows out into the Indian Ocean from the east coast of Africa, was in a state of crisis. There was no effective government. Immigrants had poured into the country at the same time that a drought occurred (*drought* means there wasn't enough rain). Crops had failed. As if that weren't bad enough, roving bands of armed thugs began terrorizing the Somali people and stealing the food they had. People, especially children, began dying of hunger. Then our marines arrived with food brought from the United States. In Somalia, as in Iraq, we worked with the United Nations. Our aim was to make peacekeeping a whole-world venture.

But neither war was as simple as it sounds. We killed innocent civilians in Iraq; lots of them. Somalia was in such disorder that its people couldn't seem to rule themselves. And we didn't want to stay there. Wars between rival religious and ethnic groups were breaking out around the globe. People were dying of starvation, while world population was zooming upward. We began asking ourselves some hard questions. Does the United States have a responsibility to try to solve the problems of other nations? Can we solve the world's problems? Or should we concentrate on creating a just society at home and hope that the rest of the world will take notice?

Those were some of the questions that faced the man who was elected president in 1992. There was something appropriate about the American people picking a leader who had been born in a town named Hope. And it was fitting that Bill Clinton had a big smile and a confident manner to go with that beautifully named hometown. He came from the center of America, and he took the opportunities the nation offered and made the most of them. His father had died before he was born; his stepfather was an alcoholic; he didn't have things easy. But Clinton had a cheerful, determined nature and he was eager to work hard and do well, which was exactly what he did. He excelled in school, got scholarships to the very best colleges, became governor of his home state of Arkansas, lost a reelection, didn't get discouraged, won the next time he tried, ran for president when no one thought he had a chance, and made it to the White House.

"Let us build an American home for the 21st century," said Bill Clinton at his inaugural celebration, "where everyone has a place at the table and not a single child is left behind." The new president from Hope—whose middle name was Jefferson, and who happened to be a Southern Baptist (in the 18th century that was a jailable offense)—had set himself a big job to accomplish. He wanted to help Americans pursue happiness.

Bill Clinton (left, with Vice President Al Gore) took office at a difficult time in a rapidly changing world; he was faced with a huge budget deficit, rising crime rates, and deeply troubled health-care and education systems. But he had a lot of energy and seemed determined to try.

Pop. 256.6 Million

In 1820, the median age of Americans was 16.7 years. In 1989, it was 32.7. *Median age* means that half the people are older and half younger than that age.

In 1920, 32 million Americans lived on farms. In 1960, the figure was 15.6 million. In 1988, only 5 million Americans were farm dwellers.

In 1989, California had 11.7 percent of the nation's population—a larger percentage than any one state since 1860, when 12.3 percent of the whole population lived in New York. California's population (29.1 million) was more than that of the 21 least populous states combined.

In 1990, almost half of all Americans lived in 37 large metropolitan areas.

In 1993, the Census Bureau reported that we were a nation of 256.6 million people.

41 A Quilt, Not a Blanket

The Perezes reunited: left to right, Victoria, Reyniel, and Lorenzo holding Alejandro.

Orestes Lorenzo Perez, a Cuban military pilot, stepped into his Soviet-built jet fighter plane and flew toward the United States. He was risking his life, but he believed the risk was worthwhile. He wanted to live in a land that was free.

Lorenzo kept his plane low, just over the water, so it would not be detected by radar. There was no way his wife, Victoria, and his sons, Reyniel and Alejandro, could go with him. "Don't worry," he told his wife. "I will come back for you."

As soon as he arrived in the United States, Orestes Lorenzo Perez tried to get his family out of Cuba. He tried all the legal means. He lobbied members of Congress; he founded an organization called Parents for Freedom —but nothing worked. His wife and children were hostages of Fidel Castro, Cuba's dictator. Castro would not let them out of the country.

> America is not like a blanket—one piece of unbroken cloth, the same color, the same texture, the same size. America is more like a quilt—many pieces, many colors, many sizes, all woven and held together by a common thread.
>
> —JESSE JACKSON,
> 20TH-CENTURY POLITICAL LEADER

In 1980, 100,000 Cubans, in boats big and small, made it to the U.S. across the Straits of Florida.

In the meantime, Lorenzo became part of the flourishing Cuban-American community. He learned to fly American planes and became a licensed American pilot. He came up with a plan, and he got a secret message to his wife. The message told of a spot on a highway near a beach. Then he borrowed a plane and called his wife on the telephone. They spoke of children's clothing, of his father, of the sunset. It was all a code. He said, "I'll send money to buy a TV."

"Already?" she asked, startled. Those code words meant he was coming the next day. He needed to know exactly when the sun set in Cuba. He asked about the children's shoe sizes. She said they were 5½ and 6½. The sun set between 5:30 and 6:30 P.M.

The next day, Victoria packed a lunch and she and the boys went to the beach. They spread out their towels and sat down. Two policemen were nearby. Victoria had brought a Bible. She read it. She tried to look like any other mother out for a day on the beach with her children. But Reyniel, who was 11, wanted to go home. He didn't want to swim. She hadn't told the boys of the plans. "Go swim," she whispered. "This is a matter of life and death." Reyniel knew something important was happening. He swam.

At five o'clock, casually, they got up to leave. At 5:45, they were standing beside the highway when they saw a plane landing two blocks away. "Run, run!" said Victoria. "It's Daddy!" The plane missed a car, a bus, a huge rock, and a traffic sign. It came to a stop about ten yards from a truck. The startled driver managed to hit his brakes just in time. The plane's pilot, Orestes Lorenzo, stayed on the ground for about 40 seconds, which was enough time to pick up his wife and boys. Then he turned the plane around and took off. He had told his wife, in the secret message, not to talk or hug him; he would need all his powers of concentration. Twenty-one minutes and forty-three seconds later, he shouted out, "We did it!" They were in United States territory. They were free.

Bang Huy Le's mother shook his bed. He got up sleepily. Two strange men were in his house. Bang Huy, who was seven years old, was soon squeezed beside his 14-year-old sister in a small open boat with 50 other Vietnamese. Out at sea, they were attacked by pirates, who took their few possessions. But they were lucky; they made it to Indonesia, and, six months later, to the United States. Four years after that, when Bang Huy had almost forgotten the Vietnamese language, his parents, his grandmother, his younger sister, and his two brothers all joined him in America.

Bang Huy Le

It remains true that nearly all of the world's richest countries are free (meaning, among other things, democratic) and nearly all of the poorest countries are not....Across the world, in other words, the correlation between political freedom and prosperity is a close one.
—THE ECONOMIST, AUGUST 1994

As the U.S. pulled out of Vietnam in 1975 and the North Vietnamese closed in on Saigon, thousands of South Vietnamese fled. Many crossed the South China Sea in tiny fishing boats like these—and some didn't make it; storms wrecked many of the boats, and others were hijacked by pirates.

In 1849, Kentucky senator Garrett Davis wrote of the dangers of immigration. "The German and Slavonic races are combining in the state of New York to elect candidates of their own blood to Congress. This is the beginning of the conflict of races on a large scale, and it must, in the nature of things, continue and increase....If it does not become a contest for bread and subsistence, wages will at least be brought down so low as to hold our native laborers and their families in hopeless poverty." Was Senator Davis right in his prediction?

The Lorenzos and Les were part of a long immigrant tradition in America. It began tens of thousands of years ago, when the first immigrants came from Asia—on foot or by dogsled or in small boats. The previous inhabitants of the land—the birds and animals—must have been surprised by the newcomers.

The people spread out over the two great American continents. Then, just 500 years ago—an eyeblink in the long view of time—new immigrants arrived from lands across the Atlantic Ocean. It was a meeting of two worlds; each had been unaware of the other.

At first, those who settled in the region that became the United States came mostly from Great Britain (England, Scotland, Wales, and Ireland), and from Africa. The Africans came unwillingly. They were forced to become workers in a society that needed them badly.

A new nation was founded—the United States of America. It was an unusual nation—a nation of free citizens (except, of course, for those who were slaves) in a world that was mostly unfree. It was born with the idea that people could govern themselves. It was a democracy.

Freedom, and the opportunities of a big, rich land, were like a magnet. People came. Lots of different people came. Many—like many of the English before them—were failures or outcasts in their old world. Comfortable people don't usually leave their homes.

Some of the English-Americans became concerned about the newcomers. Benjamin Franklin worried that the many German immigrants weren't bothering to learn English. He feared that the United States might become a two-language nation. He needn't have worried; their children quickly learned English. They became Americans. Another large group of 19th-century immigrants came from Ireland. They were very poor, but they worked and studied hard and soon they, too, were Americans.

When gold was discovered in California in 1849, boatloads of people came from all over the world—some from China and other nations in what was called the Far East (it is actually west of California).

It was astounding. No matter where they came from, everyone wanted all the freedoms and rights that were in the U.S. Constitution. They wanted to be Americans. But some of them, being human, became jealous of the next group of newly arriving immigrants.

So, when the 19th century began to turn into the 20th, and more new people came—this time speaking Italian and Polish and Russian and Greek and Turkish and Yiddish—the earlier immigrants worried. They said the newcomers would never learn English. They said they were poor people and uneducated. They said they were outcasts. All that had been said before, and much of it was true. But

Technology, Invention, DNA, and All That Stuff

Indiana's Howard Hathaway Aiken was generous. Actually, he had the same notion that Benjamin Franklin had. He thought scientific ideas should be shared. So he didn't patent his inventions—he let anyone who wanted use them. His biggest idea took shape as Mark I. Mark was really big: he was 8 feet high, 51 feet long, had 530 miles of wire in his insides, and was known as a "superbrain." Born in 1943—with IBM's money and Harvard's resources—Mark was the first completely automatic computer.

Mark used mechanical switches to do its calculations, something like those on the telephone switchboards of the day. They were reliable—and much faster than the brains of any team of human mathematicians—but still, they were slow. Electric switches would be much faster.

ENIAC had them. It was the first all-electronic computer. Which, in 1946, meant that it had thousands of vacuum tubes. ENIAC (*electronic numerical integrator and calculator*) was even bigger than Mark—it weighed 30 tons. And it was fast. J. P. Eckert and J. W. Mauchley designed it for the army.

Mark I attracted moths; a tiny moth could stop the behemoth in the middle of a calculation—its operators had to keep *debugging* it. ENIAC's vacuum tubes got hot; it had to be cooled down regularly. Both were frightfully expensive. It didn't seem likely that computers and calculators would ever reach ordinary people.

Then William Shockley invented something that was mostly *silicon*. Silicon is sand. His may be the most important invention of our times.

Shockley grew up in Palo Alto, California. His father was an engineer, his mother a surveyor. They taught him at home until he was eight. For Shockley, work and play were the same. He didn't waste time—until later in life, when he came up with some wacky, and despicable, racial theories. But that's another story.

After he got a Ph.D at MIT, Shockley went to work at Bell Telephone's research laboratory. Bell was looking for something to replace its mechanical telephone switches; some kind of electric-circuit switch. Vacuum tubes could do it, but the tubes got hot and burned out. Bell wanted to find a way to improve the vacuum tube.

But Shockley started thinking about something different, called a *semiconductor*. Semiconductors are materials that conduct electricity and insulate, too (which means they don't heat up). He found that electrical current flowing from a metal contact point, through a semiconductor to another contact point, and on to a metal base, gets strengthened in the process. Then Shockley and his colleagues came up with a device that didn't look like much—just a little piece of matter (*germanium*) with a few wires in it—but it could do everything the vacuum tube could, and more. It boosted electrical current, produced no heat, didn't burn out, eliminated miles of wiring, was very, very small, and very cheap to produce. A contest was held at Bell Labs to pick a name for the new device. The winning name was *transistor*.

When computer designers got hold of transistors—well, those giants ENIAC and Mark were dead ducks. Small is beautiful in the world of modern technology. And transistors kept getting smaller. Before long, there were pea-size chips made of silicon (which is a great semiconduc-

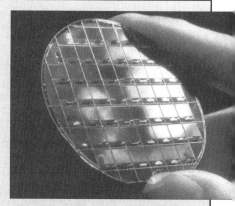

Power in a tiny space: the microchip.

tor). By the 1980s, a quarter-inch microchip held ten times the electrical components of the 30-ton ENIAC. A desktop computer could do much more than those 1940s dinosaurs.

Transistors began changing the way people lived. They were doing things that would have been considered miraculous a generation earlier. Tiny transistors in hearing aids and portable radios amplified weak signals. Transistors powered wafer-thin calculators; they were in TV sets and VCRs, in airplane control panels, in cars, microwaves, toasters, in industrial production lines, in video games, and in telephone switchboards.

Is the transistor the most important invention of our times? I kind of think so, but you can argue with me. When you use superlatives (the most, the best, the worst, etc.), it's easy to get on shaky ground.

No question, though, one of the biggest stories of the 20th century is in technology. It is a magic show that keeps going on and on. See if you can track down details of 20th-century inventive breakthroughs—especially in biotechnology. Clue: look up Francis Crick, James Watson, Hamilton Smith, and Daniel Nathans and see what they discovered.

they had forgotten that in America something happened when people were given opportunity and freedom. The new immigrants worked hard. They invented. They built. They achieved. They learned. Soon those people from southern and eastern Europe were successful Americans, and old-timers themselves.

Then, near the end of the 20th century, another group of new immigrants began arriving. Many spoke Spanish. Some people worried that they wouldn't bother to learn English and that the United States would become a two-language nation. But it wasn't likely. Their children were like all the children before them—eager to learn.

It was the same with Asians, who now arrived in force. (Until 1965, laws restricted Asian immigration.) The Asians, like Bang Huy Le and his family, were searching for opportunity and freedom. They would find it.

People came from Africa now because they wanted to be Americans. And, like all the immigrants who had arrived before them, they brought talent and energy. Other African Americans—whose ancestors had been in America for a long, long time—were finally permitted to achieve success in large numbers. Civil rights laws had opened doors to schools and jobs that for too long had been shut.

All these peoples were changing the look of America. They were making us rich in imagination and achievement.

For we all benefit from the American magic that finds wonders in the world's outcasts. There is nothing secret about that magic. It is called respect for each person's inalienable rights. It is called *freedom*.

Piecing the Quilt

IN 1820, WHICH WAS THE FIRST YEAR THE UNITED STATES kept statistics on immigration, 8,385 foreigners entered the country with the intention of becoming citizens. Most were English, Scotch-Irish, or German, but they also included 20 Danes, 14 Russians, 6 Asians, 5 Poles, and one Mexican.

THE UNITED STATES ADDS ONE PERSON TO ITS TOTAL population every 14 seconds. That adds up to about 6,300 people every day. Some 4,400 come from the greater number of births than deaths (10,600 over 6,200). The rest of the daily increase comes from immigration. We add one immigrant every 35 seconds (and one person leaves the country every three minutes). One birth occurs every eight seconds, one death every 14 seconds.

ONE-FIFTH OF THE U.S. POPULATION INCREASE IN THE 1980s came from immigration. The largest groups were Mexicans, Filipinos, Chinese, Koreans, and Vietnamese. The 1990 census counted 21 million Hispanics, a 44 percent increase since 1980. The Census Bureau predicts that by the year 2013, people of Spanish-speaking origins will be the largest minority group in the United States.

BY THE MIDDLE OF THE 21ST CENTURY, THE UNITED States is expected to have a population of 383 million that is 53 percent white (excluding Hispanics), 21 percent Hispanic, 15 percent African-American, 10 percent Asian-American, and one percent American Indian, Eskimo, and Aleut.

42 The Land That Never Has Been Yet

Douglas Wilder won a medal for bravery in the Korean War. The year he was elected governor of Virginia, David Dinkins became the first black mayor of New York City.

"As a boy, when I would read about an Abe Lincoln or a Thomas Jefferson...when I would read that all men are created equal and that they are endowed by their creator with certain inalienable rights...I knew it meant me," said Douglas Wilder, who in 1989 was elected governor of Virginia.

Virginia's capital, Richmond, had once been the capital of the Confederate States of America. The Confederate states were the slave states. Some of Wilder's ancestors were slaves. Do you think those slaves ever imagined that their great-grandson would be sitting in the governor's mansion in Richmond?

Well, maybe they did, because that idea of freedom is so powerful—especially when you don't have it—that it sets you dreaming and planning and hoping. And it is the people with dreams and hopes who make things happen.

The United States was founded by dreamers—men like Thomas Jefferson, John Adams, and Ben Franklin—who had a vision of freedom and fairness that spoke to all people, rich and poor, all over the world.

But that vision got cloudy. Some people weren't treated fairly. Some other people weren't ready for a nation where every person was to be free and equal.

A poet named Langston Hughes had something to say about that:

Langston Hughes

Until recent decades, most people worked with their hands—not their heads. Reading wasn't essential for most work. If you were a farmer, carpenter, or factory worker, you learned work skills on the job. But in the 21st century, to succeed, almost everyone—including the farmer—will need to be a thinker.

A poet who was farseeing (as poets are supposed to be) wrote in 1916: *Things fall apart; the center cannot hold; / Mere anarchy is loosed upon the world.* (His name was William Butler Yeats.) What did he mean?

Clue: one historian says that we are in the midst of a change—from an industrial society to an information society—that is as big as the Neolithic revolution, 10,000 years ago. That's when we went from being hunter-gatherers to being farmers and city dwellers who lived in smaller families and belonged to religious and political groups (city-states, nations, etc.).

But our families, religions, and cities are in the turmoil of change. Will things fall apart? Or will we make the most of the knowledge revolution?

197

1929–1968

"A man dies," said Martin Luther King, Jr., in 1965, **"when he refuses to stand up for what is right, for what is justice, for what is true."**

*Oh, let America be America again
The land that never has been yet—
And yet must be—the land where every man is free.
The land that's mine—the poor man's, Indian's, Negro's, ME—
Whose sweat and blood, whose faith and pain
Must bring back our mighty dream again.*

"Must bring back our mighty dream again." Would it happen? Could America be what it was meant to be?

*Oh yes,
I say it plain,
America never was America to me.
And yet, I swear this oath,
America will be!*

The astonishing thing about the 20th century was that it was a time when we listened, at last, to dreamers and poets like Langston Hughes.

Yes, despite all the troubles of the 20th century, some great things happened. For the first time in our history, the United States began to be a nation for all its people. Segregation—and its terrible unfairness—was finally thrown out. Some extraordinary individuals led us back to the Founding Ideas: one was a minister from Georgia named Martin Luther King, Jr.; another was a Supreme Court justice from Maryland named Thurgood Marshall. But, mostly, it was ordinary citizens, like Rosa Parks, who marched and voted and hoped and dreamed—and made things happen.

When Alfred E. Smith ran for president in 1928, he was defeated because of his Catholic religion. But, in the second half of the century, Catholic John F. Kennedy became president. Finally, people who had once been on the outside of American society—blacks, Jews, Asians, Hispanics, Native Americans, Catholics, and women—were sitting in seats of honor and prestige. It was an astounding shift—achieved mostly through nonviolent methods and changes of law.

But there was another side to the 20th century: violence—war, terrorism, and crime—had marked the times. Now gangster nations and gangster individuals, with easy-to-get bombs and guns, were making people live in fear.

We were the world's richest nation—did that give us special responsibilities? Were we meant to be global cops?

Technology was something else to ponder: would we control it, or would it control us?

Much was promising. Doctors were going to be able to prevent future illnesses by changing the DNA of babies still inside a mother's belly. Computers were being made intelligent; no one could foresee just how bright they would be, but it was clear that many of the tasks that people had done in the past—especially jobs no one really likes to do—could be handled faster and better with electronics. But that promise was also bringing a problem: would it leave masses of people with no useful place in society? Or will time become a gift that we can all use to pursue happiness?

And what about our system of education? The information age demands thinking citizens—not rote memorizers. That means new kinds of schooling. Are you being educated to be an informed democratic citizen? Are you being educated to be responsible—for yourself and others?

If so, in the 21st century you can help America become what it was meant to be…

The land that never has been yet—
And yet must be—the land where
every man is free.

"Liberty without learning is always in peril," said John F. Kennedy, "and learning without liberty is always in vain."

A National Sickness

In September 1994, an 81-year-old lady was attacked and robbed in her apartment. People gasped when they heard her name. It was Rosa Parks, and, once again, she made people aware of an enormous national sickness. This time the malady was crime, and it affected everyone.

In the 1950s (the time when Rosa Parks was protesting against bus segregation), parents in Alabama didn't usually worry when they sent their children out to play. Boys and girls walked to school, walked to their friends' houses, and walked to the store—on their own.

By 1994, many parents had to take their children everywhere. They were scared that something might happen to them if they were out alone. Ten-year-olds were carrying guns, and killing people. Almost nowhere did people feel completely safe. Crime had become the leading concern for all of America's citizens. It was a special concern for young people; nearly one in four of the victims of violent crime was now a juvenile (under 18). Among the crimes in 1994: a six-year-old was killed in gun crossfire, a teacher was shot in school, and an elderly woman, who had been a national heroine, found she was no longer safe at home. Does crime, and the fear of crime, affect you? How? What can we do about it?

Chronology of Events

1945: Harry S. Truman becomes 33rd president on the death of Franklin D. Roosevelt

1945: Japan surrenders; World War II is over

1947: President Truman launches the Marshall Plan, giving economic aid to war-crippled European countries

1947: the Truman Doctrine commits the U.S. to helping any country fighting communism

1947: first baseman Jackie Robinson joins the Brooklyn Dodgers, ending segregation in major-league baseball

1948: Truman defeats Republican Thomas E. Dewey

1950: Senator Joseph McCarthy stirs up anti-communist hysteria with a nationwide witch hunt

1950: after a surprise attack by communist-run North Korea on South Korea, an American-led United Nations force enters the Korean War

1952: World War II general Dwight D. Eisenhower elected 34th president, the first Republican in 24 years

1953: the Korean War ends; Korea remains divided in two

1954: in *Brown* v. *Board of Education*, the Supreme Court finds "separate but equal" unconstitutional in schools

1954: France pulls out of French Indochina (Vietnam); the U.S. sends in a small number of military "advisers"

1955: Dr. Martin Luther King, Jr., leads a successful black boycott of segregated city buses in Montgomery, Alabama, in response to the arrest of Rosa Parks

1957: Eisenhower sends federal troops to enforce high school desegregation in Little Rock, Arkansas

1959: Alaska and Hawaii become 49th and 50th states

1960: Democrat John F. Kennedy, a Roman Catholic, becomes the youngest elected president

1961: President Kennedy founds the Peace Corps to send volunteers and technical aid to developing countries

1962: African-American students and teachers across the South stage anti-segregation sit-ins at lunch counters

1962: in the Cuban missile crisis, JFK refuses to let the Soviet Union install nuclear missiles in Cuba

1963: in Birmingham, Alabama, Dr. Martin Luther King, Jr., and many others, including children, are jailed for demonstrating against segregation

1963: around 250,000 people march on Washington, D.C.; Dr. King delivers his "I have a dream" speech

1963: the U.S. has 11,000 military "advisers" in Vietnam

1963: the U.S. and U.S.S.R. sign Nuclear Test Ban Treaty

1963: on November 22, President Kennedy is assassinated in Dallas; Lyndon Johnson becomes 36th president

1964: the Civil Rights Act outlaws discrimination in employment and public places; Johnson is elected president by the biggest majority ever

1964: Dr. King receives the Nobel Peace Prize

1965: the Voting Rights Act outlaws practices designed to prevent black people from voting

1965: President Johnson launches his Great Society anti-poverty programs such as Medicaid and AFDC

1967: opposition to the Vietnam War soars; in Detroit, 38 die in race riots and looting; the Black Power movement gathers strength

1968: LBJ announces that he will not run for re-election

1968: Martin Luther King, Jr., is assassinated by James Earl Ray, an escaped convict who is later captured

1968: Robert F. Kennedy is assassinated while campaigning for the presidency in California

1968: Republican Richard Nixon elected 37th president; he begins by withdrawing some troops from Vietnam but later expands the war into Cambodia

1969: the space program lands a man on the moon

1972: Nixon is the first president to visit communist China; he is reelected president by a large majority

1973: North and South Vietnam sign a peace agreement

1973: Vice President Spiro Agnew resigns on charges of tax evasion; Gerald Ford appointed successor. Investigation of the Watergate scandal begins, and, to avoid being impeached, Nixon becomes first president to resign. Ford becomes 38th president and pardons Nixon

1976: Jimmy Carter defeats Ford; he is 39th president

1979: inflation hits double digits, and a Middle East oil embargo creates an energy crisis in the U.S.

1979: Muslim fundamentalists seize power in Iran and keep 52 Americans hostage for more than a year

1980: Republican Ronald Reagan defeats Carter to become 40th president; the Iranian hostages are freed

1981: the national debt passes $1 trillion

1985: Mikhail Gorbachev, the new Soviet leader, begins the process that leads to democratization of the U.S.S.R.

1988: Vice President George Bush elected 41st president

1989: a series of "peaceful revolutions" frees eastern Europe from Soviet domination; the Cold War is over

1990: Iraqi president Saddam Hussein invades Kuwait; U.S.–led international troops drive him out in the Gulf War

1992: Democrat William (Bill) Clinton defeats Bush to become the 42nd president

More Books to Read: *Referring* to Books of Reference

This may sound strange to you, but I like to read dictionaries. Now, for bedtime reading I'll take a novel, or a biography, or a good history book—but for browsing, I sometimes choose reference books.

The other day, I was strolling in the AMERICAN HERITAGE DICTIONARY OF THE ENGLISH LANGUAGE, which not only has word definitions but also has word histories and interesting notes on how to use words. I found that the word *nerd* first appeared in the language in 1950, in Dr. Seuss's book *If I Ran the Zoo*. The dictionary's editors describe a nerd as "a small humanoid creature looking comically angry, like a thin, cross Chester A. Arthur." (Who is Chester Arthur? Read Book 8 in this series and find out.) The *American Heritage Dictionary* is big and expensive, so you may have to use it in your library's reference section. THE HARCOURT BRACE STUDENT DICTIONARY is more affordable. Its editors are not only good *lexicographers* (what is that?), they have wit. This dictionary is filled with literary quotations, short essays on good writing, and clear definitions.

If you want to back up your thoughts when you're writing a paper, try quoting a famous person. In your library you should find a variety of dictionaries of quotations. The two most famous are BARTLETT'S FAMILIAR QUOTATIONS and the OXFORD DICTIONARY OF QUOTATIONS. In the *Oxford* I found this about economics from Harry Truman: "It's a recession when your neighbour loses his job; it's a depression when you lose yours." (Note the spelling of *neighbor*. The *Oxford* was published in England, and the English have *neighbours*.)

I hope you're a history buff. If so, I recommend that you *peruse* the YOUNG READER'S COMPANION TO AMERICAN HISTORY (ed. John A. Garraty, Houghton Mifflin, 1994). You use it to look up people, events, places, and terms, and get short, lively descriptions.

What does *peruse* (pur-OOZ) mean? In the OXFORD ENGLISH DICTIONARY, you'll learn that it means *to examine* or *study*. And you'll also learn that

the word first made its way into print in 1578, in a book that talked about "The diligent peruse of this History of Bones." Now the *Oxford English Dictionary* is really big (if you're in the know, you'll call it the *OED*). It is 20 volumes long, defines more than half a million words, and includes some 2.4 million quotations. Most people think it is just for scholars, but you may enjoy browsing in it, too. The *OED*, like many reference books, is now on CD-ROM.

The reference book that I use most often is ROGET'S INTERNATIONAL THESAURUS. A *thesaurus* (thuh-SOR-us) is a book of synonyms, and no writer I know can do without one. Finding the best way to say something usually means searching for just the right word; that's what a thesaurus helps you find.

If you really want to impress people with your vocabulary of big words, try Eugene Ehrlich's HIGHLY SELECTIVE THESAURUS FOR THE EXTRAORDINARILY LITERATE (HarperCollins, 1994). There, for the word *changeable*, you will find: *labile, mercurial, protean, tergiversatory,* and *versicolor.* For *celebration*: *callathump, callithump,* and *potlatch.* (Potlatch! If you read *The First Americans*—Book 1 of the series—you don't need a thesaurus to know about those celebrations.)

Eugene Ehrlich, by the way, is Tamara Glenny's (see page 203) father-in-law. So mentioning his book could be called *nepotism* (NEP-uh-tiz-um). That, according to one of my dictionaries, is *favoritism shown to a relative or close friend,* and it comes from the Italian *nepotismo,* which means "favoring of nephews," which was what some 16th-century Italian priests were accused of doing.

Have you had enough of reference books? I haven't even mentioned encyclopedias. One that I use is the ENCYCLOPEDIA BRITANNICA, another is the COLUMBIA ENCYCLOPEDIA, a third is Harper & Row's ENCYCLOPEDIA OF AMERICAN BIOGRAPHY, and a fourth is the ENCYCLOPEDIA OF WITCHCRAFT AND DEMONOLOGY. Witchcraft? Head for your library reference section and you may be amazed at what you'll find.

A Note From the Author

Dear Reader,

Boys and girls often ask me if writing is easy. The answer to that question is a big NO. Writing is hard work. Sometimes it is very tiresome. But when you have written something that is pretty good, you usually feel proud and pleased with yourself. And that is why I like writing. I like the feeling that comes after the hard work.

When I began *A History of US* I gave some chapters to Patti Frisch, who is married to my brother, Roger. Patti is a lawyer who asked tough questions and made me think hard about how I wanted to write these books.

Then I took chapters to Linkhorn Park School in Virginia Beach and asked boys and girls there what they thought of them. I had enough written to keep them busy for eight weeks. I didn't know what they would think, but they liked the stories so much they had a party to tell me so. I decided to keep writing.

Next I asked some experts. "What do you think of this history?" I asked Diane Ravitch, Al Shanker, Ernest Boyer, Paul Gagnon, Tom Dunthorn, and Dennis Denenberg. They are all educators who are trying to make schools do the best job possible. And they all encouraged me! That was wonderful. I might have gone back to newspaper writing or teaching (two things I also like to do), or tried something else, if they hadn't been so helpful.

But no one wanted to publish the books. I tried one publisher after another. No luck. Then I met my friend Paul Nagel (who is a great historian). He said, "I know someone who can help you. He is a terrific editor and publisher." That someone was Byron Hollinshead. Paul was right: Byron is terrific, and he did help me. To begin, he got his friend Elihu Rose to help finance the books. Since Elihu is a

historian as well as a businessman, that was especially appropriate. Then Byron guided me with ideas, encouraged me when I was discouraged, set me on track when I got off course, and, as an especially nice bonus, became a good friend.

Lots of people were eager to encourage a storytelling historian. Here are some others who kept me writing: Robert Mason, Chester Finn, Paul Grand, Ginny Jones, Arthur Woodward, Kim Dennis, Ruth Wattenberg, Liz McPike, Shirley Bueche, Carey McMillan Hemphill, and Ted Wolf.

Teachers helped—wonderful teachers and administrators—in California, in Illinois, in New York, in Colorado, and some other places, too. Many of them used the manuscripts in their classes and told me what worked and what didn't. A few of their names are: George Coggan, Barbara Boone, Glen Thomas, Rod Atkinson, Diane Brooks, K. Edwin Brown, Ben Troutman, Stacey Voigtsberger, Katherine Edmonds, Jean Cammer, Judy Doyle, Phyllis Clarke, Lillian Burt, Shirlee Rodriguez, Carole Young, Kay Karsen, Scarlett Chandler, Paula Walker-Nevels, Dorothy Ingram, Rachel Hopkins, John Tucker, Frank Giuliano, Barbara Stanwood, Lillian Brinkley, Eileen Vernon, Jody Smothers Marcello, and Kerry Jon Juntunen (who shared his knowledge of economics).

Some historians gave advice. Bernard Weisberger commented with perception and wit on the first five books. James McPherson shared his expertise on the Civil War and his wisdom as a historian. Robert Rutland, Virginius Dabney, and David McCullough are historian friends who smiled on these books, and that helped. So, too, did David Donald, who was my teacher in college, and who made me understand that history can be all about ideas as well as about people and events.

Since *A History of US* was meant for young people, I asked boys and girls to read the books and comment. Here are the names of just a few of those student editors: Rebecca Perkins, John Reed, Sara Dutra, Jesse Rees, Allen Adams, Velisha Bowden, James Beam, Jennifer King, Amy Burns, Lee Browning, Maeghan Aten, Nga Nguyen, Laura Pizoni, Alex Gregory, Beryl Gregory, Conswela White, Heather Harrison, and Vincent Bousquet. (I want to thank everyone who read and apologize for not listing all your names.)

There were other kinds of experts who had knowledge that they shared graciously: Barbara Klaw (a skilled editor), Lonn Taylor and Elizabeth Sharpe (of the National Museum of American History), Melanie Bierman (of the National Trust for Historic Preservation), Kristin Onuf (at Monticello), Elaine Reed (at the National Council for History Education), Stephen Sandell (at the Hubert H. Humphrey Forum), Tom and Sherry Vaughan in Oregon, and Patty Masterson (a teacher who answered my questions on grammar).

Organizations got involved: the American Federation of Teachers, the Olin Foundation, the Banyan Tree Foundation, and the Lounsberry Foundation. The Virginia Center for the Creative Arts granted me some productive weeks. The staff at Colonial Williamsburg answered my many questions.

Boys and girls are always asking me how I learned all the things that are in *A History of US*. Well, almost everything came from books. And I got those books at places like the Tattered Cover bookstore in Denver, the Strand bookstore in New York, the Bibliopath in Norfolk, and many university and used-book stores. I like bookstores—especially old ones with books that other people have read and enjoyed. I also like children's bookstores, where I have found people who read and love books, like Sue Lubbock at The Bookies, in Denver, Jewell Stoddard at the Cheshire Cat, in Washington, D.C., and Ellen Scott at the Bookhouse in Omaha.

When I was a girl in Rutland, Vermont, I went to the library to hear stories read aloud, as well as to get books to bring home. Now that I live in Virginia Beach, I still go to library readings, and I still get library books to bring home, but I also telephone my library to get answers to questions. The librarians find an answer to almost any question that I ask. Do your librarians answer questions? Have you asked?

Writing a book isn't enough. You need to get it published. I finally found a publisher, and an outstanding one—Oxford University Press. Oxford is the oldest publisher in the world (which, for a historian, is kind of nice). Laura Brown at Oxford had faith in my books when I was getting very discouraged. D. C. Heath, another fine old publisher, is distributing these books to schools (thanks to Kathy Shepard). Susan Buckley planned the teacher guides; Deborah Parks and Elspeth Leacock wrote them;

they are excellent. Charles Gibbs, Annie Stafford, and Katharine Smalley, at Oxford, have been cheerful and efficient in helping promote the books.

American Historical Publications (with Byron Hollinshead in charge) produced these books. Tamara Glenny was my editor. A lot of what the books became is thanks to Tamara. She also wrote the picture captions in the books and chose the pictures along with Mary Blair Dunton, who was the picture researcher; everyone loves the pictures, and so do I. Mervyn Clay designed the books, Elspeth Leacock planned the maps, and Wendy Frost drew them (you can see for yourself how talented they all are).

When I needed computer help (often), I called on friends: Jon Geissinger, Norman Cohen, BJ Leiderman, and Jerry Rowe. BJ (who spends most of his time composing music) also took the photo for the jacket on some of these books.

Hold on, I haven't finished. Edna Shalev did research. My children—Todd Johnson, Ellen Johnson, Jeff Hakim, Hanna Sandler, and Daniel Hakim—read chapters and listened to my problems, and that helped, too. Danny spent a summer as my research assistant. My husband, Sam, put up with my moods, cooked a lot of dinners, read and commented on chapters, and was always there when I needed him.

Some children inspired me to try to do my best. A few of their names are: Natalie and Sam Johnson; Jessica Frisch; Rachel, Zachary, and Sara Levine; Molly and Mitchell Phillips; Sophie Brown; Alexander Adler; Bang Huy (John), Diem (Linda), Quoc, and Chau Le; Alex and Jack Bethel; Will and Matt Johnson; Sam, Mickey, and Margie Ehrlich; and Page and Victor Hakim. I hope that they, and you, will begin exploring history in these books. I hope you will keep on reading until you are 100 (or older). As you journey through the 21st century, it will help a lot to know about the people who came before you.

Your friend,

Virginia Beach, Virginia
January 1995

Joy Hakim

Picture Credits

HST: Harry S. Truman Library, Kansas City, MO
JFK: John F. Kennedy Library, Boston, MA
LBJ: Lyndon B. Johnson Library, Austin, TX
LOC: Library of Congress

NA: National Archives
NBL: National Baseball League
NPS: U.S. National Park Service
NYPL PC: New York Public Library Picture Collection

Schomburg: Schomburg Center for Research in
   Black Culture, New York Public Library
UPI / B: United Press International / Bettmann Archive
WW: Wide World Photos

Cover: Ralph Fasanella, *Night Game (Practice Time)*, 1979. From the collection of the artist; **5**: WW; **6 (top left)**: NYPLPC; **6–7**: State Historical Society of Wisconsin, negative # WHi (X3) 36638, CF 6797; **7 (top right)**: Paul Schutzer, *Life*; **8**: NYPLPC; **9 (top)**: Jack Lambert, *Chicago Sun-Times*; **9 (bottom)**: NA / Women's Bureau; **10**: Magnum Photos; **11**: Philadelphia Convention and Visitors Bureau; **12**: Roy Justis, *Minnesota Star*, 1947; **13 (top)**: LOC; **13 (bottom)**: UPI / B; **14 (top)**: LOC; **14 (bottom)**: HST; **15 (left)**: Truman family; **15 (right)**, **16 (top and bottom)**: HST; **17**: UPI; **18**: NBL; **19**: William Gladstone; **20 (top)**: NBL; **20 (bottom)**: *Los Angeles Dodgers, Inc.*, Maurice Terrell; **21 (left, top right)**, **21 (bottom right)**: WW; **22**: Cleveland Public Library; **23**: UPI; **24 (left)**: Hoover Institution, Stanford, California; **25**: UPI; **26 (top)**: UPI / B; **26 (bottom left)**: *Arkansas Gazette*, June 14, 1953; **26 (bottom right)**: Charles G. Werner, *Indianapolis Star*, 1949; **27**: *Kansas City Star*, 1949; **29 (left)**: Eric Lessing / Magnum; **29 (right)**: Burt R. Thomas, *Detroit News*, 1949; **30 (top)**: WW; **30 (bottom)**: *Krokodil*; **31**: UPI / B; **32**: Museum of Modern Art, New York, gift of the Congress of Industrial Organization; **33**: Abbie Rowe / NPS; **34**: UPI / B; **35**: Abbie Rowe / NPS, © HST; **36**: UPI / B; **37 (top)**: *St. Louis Globe-Democrat*; **38**: WW; **38 (top left)**: Elliott Erwitt / Magnum; **39 (top right, bottom left)**: UPI / B; **39 (bottom right)**: *New York Daily News*; **40**: Henri Cartier-Bresson / Magnum; **41 (top and bottom)**: NA; **42 (top)**: © Al Hirshfeld, Margo Feiden Galleries, Ltd.; **42 (bottom)**: UPI / B; **43 (left)**: Francis Miller, *Life*; **43 (right)**: Leonard McCombe, *Life*; **44**: WW; **45 (top and bottom)**: UPI / B; **46**: NYPLPC; **47 (top left and bottom right)**: UPI / B; **47 (bottom left)**: NYPLPC; **48 (left)**: NYPLPC; **48 (right)**, **49 (top and bottom)**: UPI / B; **50 (top)**: Herblock © 1957, *Washington Post*; **50 (bottom)**, **51 (left)**: UPI / B; **51 (right)**: NA; **52 (top)**: *St. Louis Post-Dispatch*; **52 (bottom inset)**: Bernard Hoffman, *Life*, © 1950 Time-Warner, Inc.; **52 (bottom right)**: Margaret Bourke-White, *Life*, © Time-Warner, Inc.; **53**: UPI / B; **54 (top)**: PIX, © Time-Warner, Inc.; **54 (bottom)**: Bruce Wrighton © 1987; **55 (top)**: Holiday Inns, Inc.; **55 (bottom)**: Allan Grant; **56**: UPI / B; **57**: © Curtis Publishing Co.; **58 (top)**: Ngo Vinh Long Collection; **58 (bottom)**: War Resisters League; **59, 61 (top)**: U.S. Navy; **61 (bottom)**: Ngo Vinh Long Collection; **62**: Moorland-Spingarn Research Center, Howard University; **63 (top)**: Charles Moore / Black Star; **63 (bottom)**: LOC; **64, 65 (all)**: Alaska State Library and Archives, Juneau; **66**: photo by Alex Rivera; **67 (top)**: Carl Iwasaki, *Life*; **67 (bottom)**: NYPLPC; **68**: Ben Shahn, *Integration, Supreme Court*, 1963, Des Moines Art Center, Edmundson Art Foundation, Inc.; **69 (top)**: NYPLPC; **69 (bottom)**: © Karsh from Rapho Guillumette; **70 (top)**: *New York Times*, December 24, 1961; **70 (bottom)**: WW; **72 (top)**: NYPLPC; **72 (bottom)**: WW; **73–75 (all)**: NYPLPC; **76, 77 (top)**: Schomburg; **77 (bottom)**: Marshall Rumbaugh, *Rosa Parks*, National Portrait Gallery, Smithsonian Institution, Washington, D.C.; **78 (top, inset)**: Schomburg; **78 (bottom)**: NYPLPC; **79**: Charles Moore / Black Star; **80 (top)**: Schomburg; **80 (bottom)**: UPI /B; **81, 82 (top)**: NYPLPC; **82 (bottom)**: Schomburg; **83 (top)**: UPI / B; **83 (bottom)**: Burt Glinn / Black Star; **84 (left)**: UPI / B; **84 (right)**: NYPLPC; **85 (top)**: LOC; **85 (bottom)**: Schomburg; **86 (top)**: UPI / B; **86 (bottom)**, **87 (top and bottom)**: JFK; **87 (bottom)**: Tor Eigeland / Black Star; **89**: Alfred Eisenstaedt, *Life*; **90 (top)**: UPI / B; **90 (bottom)**: Joe Scherschel, *Life*, © Time, Inc.; **91, 93 (top)**: WW; **93 (bottom)**: Richard Yardley, *Baltimore Sun*, 1962; **94 (top, bottom right)**: WW; **94 (bottom left)**: *Prensa Latina*; **95 (top)**: UPI; **95 (bottom left)**: UPI / B; **95 (bottom right)**: Fred Powledge; **96 (top)**: State Historical Society of Wisconsin, negative # WHi (X3) 36638, CF 6797; **96 (bottom)**: Charles Moore / Black Star; **97**: WW; **98**: UPI / B; **99 (top)**: NYPLPC; **99 (top right)**: LOC, *U.S. News & World Report* Collection; **100 (top)**: UPI / B; **100 (bottom)**: Fred Powledge; **101 (top)**: A. Philip Randolph Institute; **101 (bottom)**: Federal Bureau of Investigation; **102 (top)**: Leonard Freed, 1963; **102 (middle)**: NYPLPC; **102 (bottom)**: WW;

103 (top): JFK; **103 (bottom), 104, 105 (top)**: UPI / B; **105 (bottom)**: WW; **106 (top left)**: Elliott Erwitt / Magnum; **106 (top right)**: UPI / B; **106 (middle)**: *Dallas Times Herald*, November 25, 1963; **106 (bottom)**: WW; **107 (left)**: WW; **107 (right)**: LOC; **108 (top)**: LBJ; **108 (bottom)**: LOC; **109**: LBJ; **110 (top and bottom), 111, 112 (top, bottom right)**: LBJ; **112 (bottom inset)**: LBJ National Historic Park, Johnson City; **113**: UPI / B; **114**: Karl Hubenthal; **115**: Cecil Stoughton / LBJ; **116 (top)**: G. B. Crocket, *Washington Star* / LOC; **116 (bottom)**: LBJ; **117 (top left)**: UPI / B; **117 (top right)**: NEA, Inc.; **117 (bottom)**: LOC; **118**: UPI / B; **119 (top)**: Ed Hollander; **119 (bottom)**: Bruce Davidson / Magnum; **120**: James H. Karales, *Look*, © 1965; **121 (top)**: UPI / B; **121 (bottom)**: NYPLPC; **122**: photo by Tom Lankford, *Birmingham News*; **123**: Schomburg, Lawrence Henry Collection; **125 (top, middle)**: photo by Tom Lankford, *Birmingham News*; **125 (bottom)**: Schomburg, Lawrence Henry Collection; **126**: David Levine, © *New York Review of Books*, 1971; **127 (top)**: Philip Jones Griffiths / Magnum; **127 (middle)**: Eddie Adams / AP/WW; **127 (bottom)**: Larry Burrows, *Life*; **128**: Bill Canfield, *Newark Star-Ledger*, 1968; **129**: Joy Hakim; **130 (top)**: UPI / B; **130 (bottom)**: photo by Ben Fernandez, Marquette University Memorial Library, Dept. of Special Collections and University Archives, Milwaukee, Wisconsin; **131 (top)**: David Levine © *New York Review of Books*, 1966; **131 (bottom)**: Paul Conklin, *Time*; **132 (top left, middle left)**: UPI / B; **132 (top right)**: *Detroit News*, 1967; **132 (bottom left)**: WW; **132 (bottom inset)**: NYPLPC; **133 (top)**: Yoichi R. Okamoto / LBJ; **133 (bottom), 134 (top)**: UPI / B; **134 (bottom)**: Ron Haeberle, *Life*, © Time-Warner, Inc.; **135 (top)**: *Gaslight*; **135 (bottom), 136 (right, top and bottom)**: NYPLPC; **137 (top)**, **138 (left)**: © Time-Warner, Inc.; **138 (right)**: Robert R. McElroy / *Newsweek*; **139**: Diana Mara Henry / *Viva*, 1978; **140 (top)**: LOC; **140 (bottom)**: Diana Mara Henry / *Viva*, 1978; **141**: NYPLPC; **142**: U.S. Dept. of Housing and Urban Development; **143 (top)**: Schomburg; **143 (bottom)**: *Life*, March 11, 1957; **145**: Dorothea Lange, 1936; **146**: George Ballis, 1961; **147**: WW; **149 (top)**: Joy Hakim; **149 (bottom)**: NYPLPC; **150 (top)**: Dorothea Lange, 1935; **150 (bottom left)**: Dorothea Lange, 1960; **150 (bottom right)**: Paul Fusco / Magnum; **151**: Herbert Orth; **153 (top)**: Robert Knudsen / LBJ; **153 (bottom)**: Steve Shapiro / Black Star; **154**: Georg Gerster / Ralpho / Photo Researchers; **155 (top and bottom)**: NYPLPC; **156**: UPI / B; **157**: NYPLPC; **158**: Costa Manos, © 1968; **159**: NYPLPC; **160 (top)**: photo by Gordon Parks / LOC; **160 (bottom)**: NYPLPC; **161 (top left)**: photo by Gordon Parks / LOC; **161 (top left)**: UPI / B; **161 (middle left)**: Jeff Reinking; **161 (middle middle)**: Alfred A. Knopf, Inc.; **161 (middle right)**: © Susan Mullally Well; **162 (top and bottom)**: UPI / B; **163 (top)**: Lionel Martinez; **163 (bottom)**: Bonnie M. Freer / Photo Trends; **164 (top, bottom left and right)**: UPI / B; **165 (top right)**: *New York Daily News*; **165 (bottom)**: LOC; **166 (left)**: Michael Ochs Collection; **166 (right)**: *New York Daily News*; **167 (all)**: Michael Ochs Collection; **168**: LOC; **169 (left)**: Raymond Depardon / Magnum; **169 (right)**: Cummings-Prentiss Studios; **170**: Don McCullin, *Sunday Times*; **171 (top)**: UPI / B; **171 (bottom)**: Joy Hakim; **172**: LOC; **173 (top)**: Oliphant © 1974 Universal Press Syndicate; **173 (bottom)**: UPI / B; **174**: NASA; **176**: Yoichi Okamoto / LBJ; **177 (top and bottom), 178 (middle), 179 (bottom)**: George Tames; **178 (top right)**: LOC; **178 (bottom right)**: Christopher Springmann / Black Star; **181**: UPI / B; **182**: Paul Fusco / Magnum; **183 (top)**: © Mazzotta / Rothco; **183 (bottom)**: UPI / B; **184**: LOC; **185 (top)**: Fairfax County Public Library; **185 (bottom)**: UPI / B; **187 (top)**: Beneficial Management; **187 (bottom)**: Bart Bartholomew; **188**: WW; **189**: Ira Wyman / Sygma; **190**: WW; **191**: UPI / B; **192 (top)**: UPI / B; **192 (bottom)**: Gamma / Liaison; **193**: Joy Hakim; **194**: Christopher Morris / Black Star; **195**: Dan McCoy / Rainbow; **197 (top)**: Joanne Pinneo / Black Star; **197 (bottom)**: LOC; **198**: Donna Binder / Impact Visuals; **199**: George Tames

John Travolta gets down in Saturday Night Fever (1977).

Index